Urban and Peri-Urban Forests—Status, Ecosystem Services, and Future Perspectives

Urban and Peri-Urban Forests—Status, Ecosystem Services, and Future Perspectives

Guest Editor

Miglena Zhiyanski

MDPI

Basel • Beijing • Wuhan • Barcelona • Belgrade • Novi Sad • Cluj • Manchester

Guest Editor
Miglena Zhiyanski
Forest Research Institute
Bulgarian Academy of
Sciences
Sofia
Bulgaria

Editorial Office
MDPI AG
Grosspeteranlage 5
4052 Basel, Switzerland

This is a reprint of the Special Issue, published open access by the journal *Land* (ISSN 2073-445X), freely accessible at: https://www.mdpi.com/journal/land/special_issues/9DTM5404S0.

For citation purposes, cite each article independently as indicated on the article page online and as indicated below:

Lastname, A.A.; Lastname, B.B. Article Title. *Journal Name* **Year**, *Volume Number*, Page Range.

ISBN 978-3-7258-5171-3 (Hbk)
ISBN 978-3-7258-5172-0 (PDF)
https://doi.org/10.3390/books978-3-7258-5172-0

Contents

About the Editor

Miglena Zhiyanski

Prof. Dr. Miglena Zhiyanski is a researcher and expert in forest ecology, land degradation, and ecosystem services, with a focus on the sustainable management of forest and urban landscapes. Her scientific work explores the interactions between global change drivers and forest ecosystems, emphasizing soil and vegetation dynamics, carbon storage, biodiversity, and nature-based solutions. She has coordinated and contributed to numerous international and national research projects, with results published in peer-reviewed journals and applied in practice for ecosystem restoration and urban green infrastructure planning.

As the Guest Editor of this Special Issue, she brings her extensive expertise in peri-urban and urban forest ecosystems, aiming to highlight innovative approaches that bridge science, policy, and practice to support resilient and sustainable landscapes.

Preface

This Reprint examines how urban and peri-urban forests mitigate local climates and provide essential ecosystem services. Its scope spans forest conditions, spatial patterns, management practices, and the capacity of trees and green infrastructures to respond to climate extremes, atmospheric pollutants, and other anthropogenic pressures across diverse urban contexts.

The aim is to collate recent advances that measure and value ecosystem services, monitor ecological functioning, and test management concepts and tools. Contributions range from assessments of carbon sequestration near industrial areas, the social relevance of ecosystem services in urban green infrastructures, and citizen science approaches, to plant ecophysiology, the applications of guidelines for green space management, analyses of the connectivity between natural forests and plantations, evaluations of the impacts of national parks on local livelihoods, and the use of digital geospatial twins to inform urban park design. The purpose of this Reprint is to provide evidence that supports nature-based solutions and practical decision making.

The motivation for this Reprint is the urgent need for robust, transferable knowledge that will help cities adapt to global change while maintaining the protection of human wellbeing and biodiversity. By integrating biophysical indicators, social valuation, and planning tools, this Reprint emphasizes actionable pathways to healthier, more resilient urban landscapes.

This Reprint is addressed to researchers, urban foresters, planners, landscape architects, conservation practitioners, educators, and policymakers seeking science-based guidance for planning, managing, and restoring urban and peri-urban forests.

Miglena Zhiyanski
Guest Editor

land

Article

Carbon Sequestration by Native Tree Species around the Industrial Areas of Southern Punjab, Pakistan

Muhammad Zubair [1], Ghulam Yasin [1], Sehrish Khan Qazlbash [1], Ahsan Ul Haq [2], Akash Jamil [1], Muhammad Yaseen [3], Shafeeq Ur Rahman [4,5,*] and Wei Guo [6,*]

1 Department of Forestry and Range Management, Bahauddin Zakariya University, Multan 66000, Pakistan
2 Department of Forestry & Range Management, Faculty of Agriculture, University of Agriculture, Faisalabad 38000, Pakistan
3 College of Tropical Agriculture and Forestry, Hainan University, Haikou 570228, China
4 School of Environment and Civil Engineering, Dongguan University of Technology, Dongguan 523808, China
5 MOE Laboratory for Earth Surface Processes, College of Urban and Environmental Sciences, Peking University, Beijing 100871, China
6 Farmland Irrigation Research Institute, Chinese Academy of Agricultural Sciences, Xinxiang 453003, China
* Correspondence: malikshafeeq1559@gmail.com (S.U.R.); guowei1124@163.com (W.G.)

Abstract: Industries have been a major culprit in increasing carbonaceous emissions and major contributors to global warming over the past decades. Factories in the urban periphery tend to warm cities more as compared with rural surroundings. Recently, nature-based solutions have been promoted to provide solutions related to climate adaptations and mitigation issues and challenges. Among these solutions, urban trees have proven to be an effective solution to remove air pollutants and mitigate air pollution specifically caused by carbon emissions. This work was designed to assess the role of tree species in mitigating air emissions of carbon around the vicinity of various industrial sites. For this purpose, three different industrial sites (weaving, brick kiln, and cosmetic) were selected to collect data. Selected industrial sites were divided into two areas, i.e., (a) area inside the industry and (b) area outside the industry. The samples were collected from 100 square meters inside the industries and 100 square meters outside the industries. Five different trees species comprised of four replications were selected for sampling. About twenty trees species from inside and outside of the industries were measured, making it 120 trees from all three selected industries for estimating aboveground and belowground biomass, showing their carbon estimation. The results showed that *Moringa oleifera* depicted overall higher total biomass from both inside (2.58, 0.56, and 4.57 Mg ha^{-1}) and outside sites from all three selected industries. In terms of total carbon stock and carbon sequestration inside the industry sites, *Syzygium cumini* had the most dominant values in the weaving industry (2.82 and 10.32 Mg ha^{-1}) and brick kiln (3.78 and 13.5 Mg ha^{-1}), while in the cosmetic industry sites, *Eucalyptus camaldulensis* depicted higher carbon, stock, and sequestration values (7.83 and 28.70 Mg ha^{-1}). In comparison, the sites outside the industries' vicinity depicted overall lower carbon, stock, and sequestration values. The most dominant tree inside came out to be *Dalbergia sisso* (0.97 and 3.54 Mg ha^{-1}) in the weaving industry sites, having higher values of carbon stock and carbon sequestration. *Moringa oliefra* (1.26 and 4.63) depicted dominant values in brick kiln sites, while in the cosmetic industry, *Vachellia nilotica* (2.51 and 9.19 Mg ha^{-1}) displayed maximum values as compared with other species. The findings regarding belowground biomass and carbon storage indicate that the amount of soil carbon decreased with the increase in depth; higher soil carbon stock values were depicted at a 0–20 cm depth inside and outside the industries. The study concludes that forest tree species present inside and outside the vicinity of various industries have strong potential in mitigating air emissions.

Keywords: air pollution; carbon sequestration; indigenous trees; industry

1. Introduction

The global rise of anthropogenic emissions of greenhouse gasses (GHG), especially carbon dioxide (CO_2), for about a century now is increasing in rates that have never been observed before [1,2]. This unprecedentedly increasing concentration in CO_2 in the post-industrial era could result in a rise in Earth temperatures from 2 °C to 6 °C by the end of the 21st century [2,3]. Generally, industries are the major culprit in increasing carbonaceous emissions and major contributors to global warming over the past decades [4]. Industries and factories installed in the urban periphery tend to warm the cities more than their rural surroundings [3]. Human activities such as increased factories, buildings, cars, and fewer trees enhance Urban Heat Islands (UHIs) [5]. The discharge of excessive amounts of pollutants from the industries of the developing countries warm the cities and bring severe damage to human cardiovascular and respiratory systems, thus accumulating hospital admissions and even premature mortality [6]. Studies regarding the air quality of Pakistan's major cities have depicted a continuously deteriorating air quality due to unplanned industrial installments, high population, and vehicular discharge [6–8]. The urban centers of Pakistan, such as Lahore, Multan, and Rawalpindi, have excessive CO_2, NO_x, CO, and PM_{10} [7,9]. It has been depicted that increasing and unplanned industries and factories in the vicinity of an increasing population are a major source of emissions [10]. According to the World Bank, the particulate pollution originating from industrial and vehicular emissions has posed serious health concerns. The problem has claimed 700 deaths among children and a whopping 22,000 premature deaths in adults [11].

This increasing carbon footprint can be controlled by limiting emissions using methods such as putting taxes on industries with higher carbon emissions and gasoline, thus providing incentives for industries to pollute less and conserve energy [12]. Some governments have also invested in comprehensive climate solutions involving the usage of renewable energy, increasing fuel efficiency, a clean and green energy economy via applying carbon tax, and reducing tropical deforestation [13]. As the importance of ecosystems is being realized globally for the sustainability of human well-being, a recent concept termed a nature-based solution (NBS) is being promoted to provide solutions related to climate adaptations and mitigation issues and challenges [14]. This program uses healthy and functional ecosystems to make a cost-effective contribution in meeting the challenges of regulating air quality, climate regulation, prevention of soil erosion, and economic and social development [15]. According to the international union for conservation of nature (IUCN), nature-based solutions are termed actions that aim to protect, manage, and restore ecosystems [16].

Among the nature-based solutions (NBS), green infrastructure, i.e., urban trees, via various research has proven to be an effective solution to remove air pollutants [17]. Urban forests are well-reputed in positively contributing towards maintaining environment quality [18]. Throughout the literature, there is a normative assertion from most environmentalists that an increase in urban forests helps mitigate pollution problems in the urban centers, but it is more cost-effective than many other approaches [19,20]. Many scientific studies have described the trees' carbon fixation capability, making them effective for ameliorating air quality by lessening CO_2 concentration [21–23]. A recent study in the USA concluded that restoring an urban environment with an average level of tree cover can remove up to 27% of air pollutants through interception and absorption of particulate matter [24]. In various studies, it was observed that urban forests had the capacity to eliminate NO_2 and O_3 via absorbing and diffusing O_3 via foliar gas exchange and dry deposition, thus decreasing the concentration of the pollutants in the atmosphere [25]. Nowak [23] observed trees in an urban scenario were absorbing pollutants from the air, thus decreasing air pollutants. Alonso et al. [26] depicted urban forests present in the pre-urban areas of Madrid, Spain to act as O_3 sinks. Furthermore, the fiscal efficacy of this method was compared with any other conventional method of mitigating air pollution; this nature-based solution was considered cost-effective [22,27].

The role of urban forests in sequestering carbon around the avenues, streets, highways, and parks has been considerably studied [28]. However, to our best understanding, the role of urban forests and their function regarding the fixation of carbon dioxide released from industries in the Pakistani context are very little understood. The present study determines the best possible forest species to sequester carbon within or around the vicinity of various industrial sites. The objectives of this study in Multan city, Punjab, Pakistan were to quantify carbon sequestration by urban trees located in industry or factory campuses and then in the trees located outside in the vicinity of the industry in the range of 100 m^2. Furthermore, it was aimed to determine the tree species with a higher ability to fix C into their biomass. The approach used by this study can be used to assess the actual and potential role of urban forests in reducing atmospheric CO_2 in Pakistan.

2. Materials and Methods

2.1. Study Location and Sampling Methodology

This study was conducted in the semi-arid district of Multan (30.181459, 71.492157), Punjab, Pakistan, i.e., Multan city (Figure 1 Map). According to the Köppen–Geiger classification, the Multan district falls within the desert climate (BWh). The average annual precipitation is 175 mm, and the average annual temperature is 25.6 °C. Three different industrial sites (weaving, brick kiln, and cosmetic) were chosen for sampling and data collection.

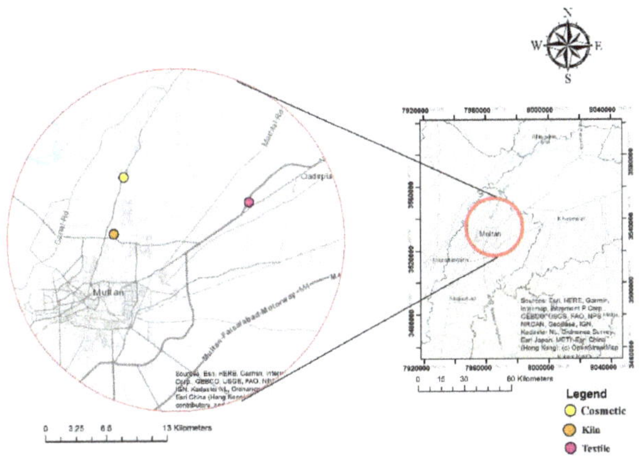

Figure 1. Map of the study site.

2.1.1. Background Information on Selected Industries

All industries selected for the study had originally been situated quite far from the city. Rapid population spread and urban sprawl brought these industries in and around the periphery of the newly built housing societies around the city. A brick kiln, one of the selected sites for the study, was just present a few kilometers away from the city. It has a large school just next to its periphery. This made for the selection of this site to understand the role of urban greening at a local scale. The cosmetic industry part of the study site was another important selection, as recently a 50-acre mango orchard in its periphery was cleared to make way for a mega housing society. While the weaving industry was part of a rural setting away from the city, it was the biggest of the lot, surrounded by mango and orange orchards.

2.1.2. Sampling Methodology

For the collection of data, the selected industrial sites were divided into two areas, i.e., (a) area inside the industry and (b) area outside the industry. The area inside the industry constituted 100 m^2. The sampled area included the industries' campus boundaries and the areas adjacent to its outside walls. The inside area also included outside parts of the industry, including the boundary of the industry, which includes roads, streets, and canals. The area outside the industry also comprised 100 square meters. The trees and soil samples were taken 100 m away from the industries' boundary walls. The sampling was conducted throughout the course of the 100 square meters in all four directions forming a square. All the industrial sites selected had native plant species present in and around their vicinity. The presence of mango orchards in all the selected study sites was common, as the area is famous for mango production. The selected tree species were dominant in the study sites, and hence were selected in the study. A total of 120 trees, 40 from each site was measured (Table 1).

Table 1. Different trees species present in different industries.

Sr. No	Weaving Industry		Brick Kiln Industry		Cosmetic Industry	
	Inside	Outside	Inside	Outside	Inside	Outside
1	*V. nilotica*	*V. nilotica*	*D. sissoo*	*D. sissoo*	*E. camaldulensis*	*E. camaldulensis*
2	*M. indica*	*M. indica*	*V. nilotica*	*V. nilotica*	*D. sissoo*	*D. sissoo*
3	*D. sissoo*	*D. sissoo*	*M. oleifera*	*M. oleifera*	*M. oleifera*	*M. oleifera*
4	*M. oleifera*	*M. oleifera*	*M. indica*	*M. indica*	*M. indica*	*M. indica*
5	*S. cumini*	*S. cumini*	*S. cumini*	*S. cumini*	*V. nilotica*	*V. nilotica*

The instrumental procedure for data collection includes the selection of trees in a 100 m radius. Each tree was then measured for height using a Suunto clinometer and girth using steel tape, whereas soil samples were extracted from different depths using soil auger.

2.2. Carbon Estimation of Above- and Belowground Biomass

For the collection of data, field visits were carried out in December 2019. Girth was measured at a height of 1.37 m from the ground level, and terminal height for each tree species within and outside the 100 square meters was measured and recorded. The tree biomass was calculated with an allometric equation from the literature (Table 2) and corrected for log bias where appropriate. The belowground biomass was assumed to be 26% of the aboveground biomass for all the tree species [29,30]. After that, individual tree biomass was converted into biomass per plot, biomass per hectare, and total carbon stock per hectare. Furthermore, we calculated the carbon content through biomass by assuming that the dry mass was 48.1% (0.48) carbon [31]. CO_2 sequestration was calculated by multiplying tree carbon with a factor of 3.66 [32].

Table 2. Different equations used to calculate the aboveground biomass of different tree species. Equations are presented transformed appropriately to calculate component biomass in kg directly. In the absence of belowground equations for a species, belowground biomass was assumed to be 26% of aboveground biomass [29].

Species	Equation	Citation
Vachellia nilotica	$10^{-1.0646} * D^2H^{0.9098}$	Ravindranath and Ostwald [33]
Mangifera indica	$2.87 * D^{0.87}$	
Dalbergia sissoo	$e^{-3.1141} * D^2H^{0.9719}$	Brown, et al. [34]
Moringa oleifera	$e^{-3.1141} * D^2H^{0.9719}$	Brown, et al. [34]
Syzygium cumini	$10^{-1.2066} * D^2H^{0.9872}$	Rai [35]
Eucalyptus camaldulensis	$e^{-2.2660} * D^{2.4663}$	Hawkins [36]

D = Diameter (cm) and H = Height (m).

2.3. Soil Sampling and Analysis

Soil data were collected at two depths, 0–20 cm and 20–40 cm, near the base of the same tree species present within and outside the 100 square meters for each industry from the four cardinal directions. Soil sampling was conducted with the help of soil auger.

Four samples were taken to make a composite sample. A total of 240 samples at both depths of 0–20 cm and 0–40 cm were collected and stored in polythene bags and analyzed at the Bahauddin Zakariya University, Multan. Bulk density was also measured by using a 100 cm^3 stainless-steel cylinder. After air-drying the sample, samples were passed through a 2 mm sieve. Organic carbon was calculated using the Walkley–Black method. After that, the value of soil depth, bulk density, and percentage of organic carbon was multiplied to calculate the carbon of soil per hectare [37].

$$\text{SOC} = \text{OC\%} \times \text{Bulk density } (\text{g cm}^{-3}) \times \text{sampling depth (cm)}$$

2.4. Statistical Analysis

Descriptive statistics were performed by using Statistics 8.1 software (Statistical package). Furthermore, one-way ANOVA, including LSD (least significant difference between means), was also applied to test the difference across all study sites.

3. Results

3.1. Growth Parameters

The results regarding growth parameters: diameter (DBH cm) and height (m) of different trees inside and outside of all the selected industrial sites: weaving industry, brick kiln, and cosmetic are represented in Table 3 ($p < 0.05$). Overall, greater diameter and height were observed for trees inside the industrial sites than outside the industries. Inside the industrial area, the estimated range of tree height and diameter was 8.15 m–19.05 m and 19.82 cm–46.92 cm, while outside the range of growth parameters were 5.03 m–14.71 m and 11.93 cm–50.96 cm, respectively. In the weaving industry, the maximum tree height (12.27 m) was computed for *Mangifera indica*, while the maximum diameter (40.85 cm) was estimated for *Moringa oleifera*. *Vachellia nilotica* depicted a greater height (14.48 m and 14.71 m) than all other trees inside and outside the brick kiln industry. Maximum diameter (40.85 cm) inside the brick kiln industry was observed for *Mangifera indica*, while outside the industry for Moringa oleifera (38.42 cm), as shown in Table 3. The maximum height inside the cosmetic, industrial area was measured for *Eucalyptus camaldulensis* (19.05 m), while the maximum value (12.57 m) was computed outside of the industry was for *Vachellia nilotica*. However, a greater diameter (50.96 cm) was estimated for *Mangifera indica* outside the cosmetic, industrial area compared with other trees (Table 3).

Table 3. Height (m) and diameter (cm) of different tree species present inside and outside the different industries. Values were reported as mean ± standard deviation.

Industries	Species	Inside			Outside		
		Height (m)	Diameter (cm)	Estimated Age Class	Height (m)	Diameter (cm)	Estimated Age Class
Weaving	*Dalbergia sissoo*	8.99 ± 1.46	20.02 ± 5.42	8–13	9.53 ± 1.92	23.05 ± 8.50	9–14
	Mangifera indica	12.27 ± 1.12	38.01 ± 17.70	8–13	9.14 ± 2.16	40.24 ± 9.94	10–15
	Moringa oleifera	12.19 ± 1.42	40.85 ± 14.66	5–10	5.03 ± 0.81	11.93 ± 0.77	1–5
	Syzygium cumini	10.97 ± 1.49	25.74 ± 15.42	8–13	9.75 ± 3.51	24.87 ± 10.00	7–12
	Vachellia nilotica	9.98 ± 1.10	32.76 ± 3.65	7–11	11.96 ± 2.98	25.48 ± 8.37	5–10
Brick kiln	*Dalbergia sissoo*	11.51 ± 1.87	21.64 ± 4.87	11–16	9.98 ± 1.45	12.34 ± 4.91	3–7
	Mangifera indica	11.28 ± 3.64	40.85 ± 9.92	5–10	10.82 ± 1.13	33.57 ± 12.78	3–7
	Moringa oleifera	9.22 ± 1.80	20.22 ± 6.60	1–5	11.74 ± 1.50	38.42 ± 23.49	7–13
	Syzygium cumini	12.80 ± 2.69	27.50 ± 17.76	8–13	9.60 ± 3.78	18.61 ± 12.61	5–10
	Vachellia nilotica	14.48 ± 1.60	27.91 ± 12.98	4–9	14.71 ± 4.51	23.46 ± 7.29	3–7
Cosmetic	*Dalbergia sissoo*	8.76 ± 1.22	22.44 ± 9.84	10–15	7.47 ± 2.73	21.64 ± 12.13	7–12
	Eucalyptus camaldulensis	19.05 ± 1.58	33.57 ± 9.98	5–10	8.38 ± 1.35	18.40 ± 2.59	1–5
	Mangifera indica	11.89 ± 3.85	43.89 ± 18.02	6–12	12.19 ± 3.59	50.96 ± 18.05	8–14
	Moringa oleifera	12.34 ± 3.08	46.92 ± 21.80	7–13	9.37 ± 1.71	37.61 ± 19.69	5–10
	Vachellia nilotica	8.15 ± 2.27	19.82 ± 9.67	3–7	12.57 ± 4.22	39.64 ± 14.67	10–15

3.2. Tree Biomass, Carbon Stock, and CO_2 Sequestration Rate

The total mean biomass showed strong variation and was significantly ($p \leq 0.05$) different among tree species present inside and outside the selected industries: weaving, brick kiln and cosmetic industry (Table 4). Overall, aboveground, belowground, and total biomass were greater for trees present inside the industrial area than outside. Inside the industrial area, the maximum total biomasses were, 2.58 Mg ha^{-1} (weaving), 2.74 Mg ha^{-1} (brick kiln), and 16.32 Mg ha^{-1} (cosmetic) for *Moringa oleifera, Syzygium cumini,* and *Eucalyptus camaldulensis,* respectively, while outside the industrial area the range of total biomass was (0.09 Mg ha^{-1}–5.22 Mg ha^{-1}) across all industries. Outside the weaving industry, the maximum total biomass was computed for *Dalbergia sissoo* (2.02 Mg ha^{-1}), while the minimum for *Moringa oleifera* (0.09 Mg ha^{-1}). Similarly, outside of the brick kiln and cosmetic industries, higher biomass was estimated for *Moringa oleifera* and *Vachellia nilotica,* whereas minimum total biomass was measured for *Mangifera indica* (0.19 Mg ha^{-1} and 0.27 Mg ha^{-1}) as depicted in Table 4.

Table 4. Biomass of different tree species present inside and outside the different industries.

Industries	Species	Inside			Outside		
		AGB Mg ha^{-1}	BGB Mg ha^{-1}	TB Mg ha^{-1}	AGB Mg ha^{-1}	BGB Mg ha^{-1}	TB Mg ha^{-1}
	Dalbergia sissoo	$0.38^{b} \pm 0.16$	$0.09^{b} \pm 0.04$	$0.48^{b} \pm 0.21$	$1.60^{a} \pm 1.21$	$0.42^{a} \pm 0.31$	$2.02^{a} \pm 1.52$
	Mangifera indica	$0.17^{b} \pm 0.07$	$0.04^{a} \pm 0.01$	$0.20^{b} \pm 0.09$	$0.18^{b} \pm 0.04$	$0.05^{b} \pm 0.01$	$0.22^{b} \pm 0.05$
Weaving	*Moringa oleifera*	$2.24^{a} \pm 1.42$	$0.58^{a} \pm 0.37$	$2.58^{b} \pm 1.79$	$0.07^{b} \pm 0.02$	$0.02^{b} \pm 0.00$	$0.09^{b} \pm 0.02$
	Syzygium cumini	$1.62^{a} \pm 0.63$	$0.42^{b} \pm 0.16$	$2.04^{a} \pm 0.79$	$1.61^{ab} \pm 0.86$	$0.30^{ab} \pm 0.22$	$1.46^{ab} \pm 1.08$
	Vachellia nilotica	$1.89^{a} \pm 0.56$	$0.49^{a} \pm 0.15$	$2.38^{a} \pm 0.70$	$1.08^{ab} \pm 0.83$	$0.28^{ab} \pm 0.22$	$1.36^{ab} \pm 1.05$
	Dalbergia sissoo	$0.57^{ab} \pm 0.24$	$0.15^{ab} \pm 0.06$	$0.72^{ab} \pm 0.31$	$0.17^{b} \pm 0.12$	$0.04^{b} \pm 0.03$	$0.21^{b} \pm 0.15$
	Mangifera indica	$0.18^{b} \pm 0.04$	$0.05^{b} \pm 0.01$	$0.22^{b} \pm 0.05$	$0.14^{b} \pm 0.05$	$0.03^{b} \pm 0.01$	$0.19^{b} \pm 0.06$
Brick Kiln	*Moringa oleifera*	$0.44^{ab} \pm 0.33$	$0.11^{ab} \pm 0.86$	$0.56^{ab} \pm 0.42$	$2.09^{a} \pm 2.16$	$0.54^{a} \pm 0.56$	$2.64^{a} \pm 2.73$
	Syzygium cumini	$2.17^{a} \pm 2.44$	$0.57^{a} \pm 0.63$	$2.74^{a} \pm 3.07$	$0.84^{ab} \pm 0.90$	$0.22^{ab} \pm 0.23$	$1.06^{ab} \pm 1.13$
	Vachellia nilotica	$2.09^{a} \pm 1.34$	$0.54^{a} \pm 0.35$	$2.64^{a} \pm 1.70$	$1.35^{ab} \pm 0.74$	$0.35^{ab} \pm 0.19$	$1.70^{ab} \pm 0.92$
	Dalbergia sissoo	$0.55^{b} \pm 0.43$	$0.14^{b} \pm 0.11$	$0.69^{b} \pm 0.54$	$0.56^{b} \pm 0.79$	$0.14^{b} \pm 0.20$	$0.71^{b} \pm 0.99$
	Eucalyptus camueldenisis	$12.95^{a} \pm 5.36$	$3.36^{a} \pm 1.39$	$16.32^{a} \pm 6.76$	$1.04^{b} \pm 0.41$	$0.27^{b} \pm 0.10$	$1.31^{b} \pm 0.52$
Cosmetic	*Mangifera indica*	$0.18^{b} \pm 0.06$	$0.04^{b} \pm 0.01$	$0.23^{b} \pm 0.08$	$0.21^{b} \pm 0.06$	$0.05^{b} \pm 0.01$	$0.27^{b} \pm 0.08$
	Moringa oleifera	$3.63^{b} \pm 3.1$	$0.94^{b} \pm 0.82$	$4.57^{b} \pm 4.02$	$1.64^{b} \pm 1.52$	$0.42^{b} \pm 0.39$	$2.06^{b} \pm 1.92$
	Vachellia nilotica	$0.76^{b} \pm 0.79$	$0.19^{b} \pm 0.20$	$0.96^{b} \pm 1.00$	$4.15^{a} \pm 3.06$	$1.07^{a} \pm 0.79$	$5.22^{a} \pm 3.85$

Means with similar letters are not statiscally different at p 0.05.

Moreover, AGB (aboveground biomass), BGB (belowground biomass), and TB (total biomass) represent the aboveground biomass, belowground biomass, and total biomass, respectively. Like tree biomass, strong variation in tree carbon stock and CO_2 sequestration was observed among various tree species inside and outside the selected industrial sites. The measured above, below, and total carbon stock of all tree species was significantly different ($p \leq 0.05$), as depicted in Table 3. Inside the weaving and brick kiln industries, the maximum total carbon stock (2.82 Mg ha^{-1} and 3.78 Mg ha^{-1}) was measured for *Syzygium cumini*; however, for the cosmetic industry, the maximum carbon stock (7.83 Mg ha^{-1}) was estimated for *Eucalyptus camaldulensis* (Table 3). *Vachellia nilotica* trees present outside the cosmetic industry exhibited greater carbon stock (2.51 Mg ha^{-1}) than trees present outside the weaving and brick kiln industries. Overall, maximum CO^2 sequestration (10.32 Mg ha^{-1} yr^{-1} (weaving), 13.85 Mg ha^{-1} yr^{-1} (brick Kiln), and 28.70 Mg ha^{-1} yr^{-1} (cosmetic) was measured for trees present inside the selected industries as compared with outside, as described in Table 5.

Table 5. Carbon of different tree species present inside and outside the different industries.

Industries	Species	Inside				Outside			
		AGC Mg ha^{-1}	BGC Mg ha^{-1}	TC Mg ha^{-1}	CO_2 Sequestration Mg ha^{-1}	AGC Mg ha^{-1}	BGC Mg ha^{-1}	TC Mg ha^{-1}	CO_2 Sequestration Mg ha^{-1}
Weaving	*Dalbergia sissoo*	0.18 ^{b} ± 0.08	0.05 ^{b} ± 0.02	0.23 ^{cd} ± 0.10	0.83 ^{cd} ± 0.36	0.77 ^{a} ± 0.58	0.20 ^{a} ± 0.15	0.97 ^{a} ± 0.72	3.54 ^{a} ± 2.67
	Mangifera indica	0.08 ^{b} ± 0.03	0.02 ^{b} ± 0.01	0.10 ^{d} ± 0.04	0.36 ^{d} ± 0.15	0.08 ^{b} ± 0.02	0.02 ^{b} ± 0.00	0.10 ^{b} ± 0.02	0.38 ^{b} ± 0.08
	Moringa oleifera	1.08 ^{a} ± 0.69	0.28 ^{a} ± 0.18	1.08 ^{bc} ± 0.69	3.95 ^{bc} ± 2.50	0.03 ^{b} ± 0.01	0.01 ^{b} ± 0.00	0.04 ^{b} ± 0.01	0.15 ^{b} ± 0.03
	Syzygium cumini	0.78 ^{a} ± 0.30	0.20 ^{a} ± 0.08	2.82 ^{a} ± 1.10	10.32 ^{a} ± 4.03	0.56 ^{ab} ± 0.41	0.14 ^{ab} ± 0.11	0.70 ^{ab} ± 0.52	2.57 ^{ab} ± 1.89
	Vachellia nilotica	0.91 ^{a} ± 0.27	0.24 ^{a} ± 0.07	1.14 ^{b} ± 0.34	4.19 ^{b} ± 1.24	0.52 ^{ab} ± 0.40	0.13 ^{ab} ± 0.10	0.65 ^{ab} ± 0.51	2.38 ^{ab} ± 1.85
Brick Kiln	*Dalbergia sissoo*	0.27 ^{ab} ± 0.12	0.07 ^{ab} ± 0.03	0.34 ^{ab} ± 0.15	1.25 ^{b} ± 0.54	0.08 ^{b} ± 0.06	0.02 ^{b} ± 0.02	0.10 ^{b} ± 0.07	0.37 ^{b} ± 0.26
	Mangifera indica	0.09 ^{b} ± 0.02	0.02 ^{b} ± 0.00	0.10 ^{b} ± 0.02	0.39 ^{b} ± 0.08	0.07 ^{b} ± 0.02	0.01 ^{b} ± 0.01	0.09 ^{b} ± 0.03	0.33 ^{b} ± 0.11
	Moringa oleifera	0.21 ^{ab} ± 0.16	0.05 ^{ab} ± 0.04	0.26 ^{ab} ± 0.20	0.98 ^{b} ± 0.73	1.00 ^{a} ± 1.04	0.26 ^{a} ± 0.27	1.26 ^{a} ± 1.31	4.63 ^{a} ± 4.80
	Syzygium cumini	1.04 ^{a} ± 1.16	1.32 ^{a} ± 1.47	3.78 ^{a} ± 4.24	13.85 ^{a} ± 15.53	0.40 ^{ab} ± 0.43	0.10 ^{ab} ± 0.11	0.50 ^{ab} ± 0.54	1.85 ^{ab} ± 1.99
	Vachellia nilotica	1.00 ^{a} ± 0.65	1.27 ^{a} ± 0.81	0.05 ^{a} ± 0.04	8.31 ^{ab} ± 5.35	0.65 ^{ab} ± 0.35	0.16 ^{ab} ± 0.09	0.82 ^{ab} ± 0.44	2.99 ^{ab} ± 1.62
Cosmetic	*Dalbergia sissoo*	0.26 ^{b} ± 0.20	0.06 ^{b} ± 0.05	0.33 ^{b} ± 0.26	1.23 ^{b} ± 0.95	0.27 ^{b} ± 0.37	0.07 ^{b} ± 0.09	0.34 ^{a} ± 0.47	1.25 ^{b} ± 1.75
	Eucalyptus camaldulensis	6.22 ^{a} ± 2.57	1.61 ^{a} ± 0.67	7.83 ^{a} ± 3.24	28.70 ^{a} ± 11.89	0.50 ^{b} ± 0.20	0.13 ^{b} ± 0.05	0.63 ^{b} ± 0.25	2.30 ^{b} ± 0.92
	Mangifera indica	0.09 ^{b} ± 0.03	0.02 ^{b} ± 0.01	0.11 ^{b} ± 0.04	0.41 ^{b} ± 0,15	0.10 ^{b} ± 0.03	0.02 ^{b} ± 0.01	0.13 ^{b} ± 0.04	0.47 ^{b} ± 0.18
	Moringa oleifera	1.74 ^{b} ± 1.53	0.45 ^{b} ± 0.39	2.19 ^{b} ± 1.93	8.04 ^{b} ± 7.07	0.78 ^{b} ± 0.73	0.20 ^{b} ± 0.19	0.99 ^{b} ± 0.92	3.63 ^{b} ± 3.37
	Vachellia nilotica	0.36 ^{b} ± 0.38	0.09 ^{b} ± 0.38	0.46 ^{b} ± 0.48	1.70 ^{b} ± 1.76	1.99 ^{a} ± 1.46	0.51 ^{a} ± 0.38	2.51 ^{a} ± 1.85	9.19 ^{a} ± 6.78

Means with similar letters are not statiscally different at *p* 0.05. Where first and second values represent mean and standard deviation. Moreover, AGC (aboveground carbon), BGC (belowground carbon).

3.3. Soil Carbon stock

The organic carbon (OC%) and soil carbon stock (SOC Mg ha^{-1}) decreased for all tree species with the increase in depth for both inside and outside of the selected industrial sites: 0–20 cm > 20–40 cm (Table 4). It was observed that overall, from both inside and outside sites, *Dalbergia sissoo* had the maximum SOC: 53.05 Mg ha^{-1}, 66.16 Mg ha^{-1} (0–20 cm), 43.11 Mg ha^{-1}, and 47.51 Mg ha^{-1} (20–40 cm). *Vachellia nilotica,* in the weaving and cosmetic industry sites, displayed the max OC (2.31%, 2.11%, and 1.9%, 1.86%) and SOC (48.16 Mg ha^{-1}, 43.08 Mg ha^{-1}, 40.53 Mg ha^{-1}, and 38.95 Mg ha^{-1}) at 0–20 cm and 20–40 cm depths, while *Moringa oleifera* showed the least values at these sites. Similarly, outside the weaving and cosmetic industry, maximum OC and SOC for both depths was measured for *Vachellia nilotica* and minimum OC: 3.08% and 2.19% (0–20 cm depth) and 2.04% and 1.95% (20–40 cm depth) and SOC: 63.16 Mg ha^{-1} and 47.07 Mg ha^{-1} (0–20 cm depth) and 42.31 Mg ha^{-1} and 40.78 Mg ha^{-1} (20–40 cm depth) for *Moringa oleifera,* as described in Table 6's first bullet.

Table 6. Soil organic carbon of different tree species present inside and outside the different industries. Where first and second values represent the mean values and standard deviation. Moreover, OC (organic carbon) and SOC (soil organic carbon) represent organic carbon and soil organic carbon.

Industries	Species	Inside						Outside					
		OC% (0–20)	SOC Mg ha^{-1} (0–20)	OC% (20–40)	SOC Mg ha^{-1} (20–40)	TOC (0–40) (%)	TSOC Mg ha^{-1} (0–40)	OC% (0–20)	SOC Mg ha^{-1} (0–20)	OC% (20–40)	SOC Mg ha^{-1} (20–40)	TOC (0–40) (%)	TSOC Mg ha^{-1} (0–40)
Weaving	*Dalbergia sissoo*	1.90 ^{b} ± 0.07	40.81 ^{b} ± 3.32	1.57 ^{b} ± 0.02	34.73 ^{b} ± 2.01	2.685 ± 0.045	58.175 ± 4.325	2.64 ^{b} ± 0.04	56.72 ^{b} ± 2.53	1.78 ^{b} ± 0.06	38.62 ^{b} ± 2.15	3.53 ± 0.07	76.03 ± 3.605
	Mangifera indica	1.49 ^{d} ± 0.03	29.43 ^{d} ± 1.11	1.32 ^{d} ± 0.02	27.17 ^{d} ± 0.38	2.15 ± 0.04	43.015 ± 1.3	2.41 ^{d} ± 0.03	47.45 ^{d} ± 1.25	1.65 ^{a} ± 0.05	33.31 ^{c} ± 1.16	3.23 ± 0.055	64.10 ± 1.83
	Moringa oleifera	1.39 ^{d} ± 0.07	28.05 ^{d} ± 0.63	1.2 ^{e} ± 0.04	24.69 ^{e} ± 0.83	1.295 ± 0.09	26.37 ± 0.73	1.94 ^{e} ± 0.05	38.20 ^{e} ± 2.53	1.39 ^{c} ± 0.03	28.54 ^{d} ± 0.77	1.66 ± 0.065	52.47 ± 2.91
	Syzygium cumini	1.75 ^{c} ± 0.08	35.53 ^{c} ± 2.38	1.46 ^{c} ± 0.07	30.46 ^{c} ± 0.87	2.48 ± 0.115	32.9 ± 2.815	2.51 ^{c} ± 0.05	52.19 ^{c} ± 1.38	1.70 ^{b} ± 0.01	35.21 ^{c} ± 1.19	3.36 ± 0.05	69.75 ± 1.97
	Vachellia nilotica	2.31 ^{a} ± 0.07	48.16 ^{a} ± 2.55	1.9 ^{a} ± 0.01	40.53 ^{a} ± 1.23	3.26 ± 0.075	68.42 ± 3.16	3.08 ^{a} ± 0.11	63.16 ^{a} ± 2.86	2.04 ^{a} ± 0.14	42.31 ^{a} ± 3.75	4.1 ± 0.18	84.31 ± 4.73
Brick kiln	*Dalbergia sissoo*	2.6 ^{a} ± 0.10	53.05 ^{a} ± 2.86	2.01 ^{a} ± 0.06	43.11 ^{a} ± 0.73	3.605 ± 0.08	74.60 ± 3.22	3.11 ^{a} ± 0.10	66.16 ^{a} ± 2.40	2.20 ^{a} ± 0.14	47.51 ^{a} ± 2.68	4.21 ± 0.17	89.915 ± 3.74
	Mangifera indica	1.51 ^{d} ± 0.12	31.38 ^{d} ± 2.26	1.32 ^{b} ± 0.04	28.19 ^{c} ± 1.93	2.17 ± 0.14	45.47 ± 3.22	2.01 ^{e} ± 0.05	41.54 ^{e} ± 1.98	1.72 ^{c} ± 0.05	36.12 ^{d} ± 1.38	2.87 ± 0.075	59.66 ± 1.68
	Moringa oleifera	2.31 ^{b} ± 0.07	49.69 ^{a} ± 1.78	1.91 ^{b} ± 0.05	42.57 ^{a} ± 0.78	3.26 ± 0.09	70.97 ± 2.17	2.88 ^{b} ± 0.08	61.99 ^{b} ± 4.29	2.00 ^{b} ± 0.07	44.08 ^{b} ± 2.10	3.88 ± 0.15	84.03 ± 6.39
	Syzygium cumini	1.93 ^{c} ± 0.03	40.40 ^{b} ± 0.15	1.72 ^{c} ± 0.02	36.82 ^{b} ± 1.05	2.79 ± 0.04	58.81 ± 0.67	2.72 ^{c} ± 0.05	55.76 ^{c} ± 2.49	1.91 ^{b} ± 0.02	39.26 ^{c} ± 0.94	3.675 ± 0.07	75.39 ± 2.96
	Vachellia nilotica	1.83 ^{c} ± 0.19	36.80 ^{c} ± 3.27	1.43 ^{d} ± 0.02	30.03 ^{c} ± 1.30	2.54 ± 0.21	51.8 ± 3.92	2.45 ^{d} ± 0.295	49.51 ^{d} ± 0.59	1.95 ^{b} ± 0.05	40.73 ^{c} ± 1.53	3.42 ± 0.17	69.87 ± 1.35
Cosmetic	*Dalbergia sissoo*	1.85 ^{b} ± 0.06	36.49 ^{b} ± 1.15	1.72 ^{c} ± 0.05	35.39 ^{b} ± 1.41	2.71 ± 0.11	54.18 ± 2.56	2.02 ^{c} ± 0.05	39.74 ^{b} ± 1.15	1.36 ^{c} ± 0.05	37.69 ^{b} ± 1.25	2.27 ± 0.075	58.58 ± 1.775
	Eucalyptus camaldulensis	2.06 ^{a} ± 0.04	42.78 ^{a} ± 1.23	1.8 ^{b} ± 0.02	38.07 ^{a} ± 1.32	2.96 ± 0.05	62.18 ± 1.89	2.19 ^{b} ± 0.03	45.56 ^{a} ± 2.06	1.89 ^{a} ± 0.06	40.01 ^{ab} ± 1.58	3.13 ± 0.06	65.5 ± 2.84
	Mangifera indica	1.46 ^{c} ± 0.06	29.42 ^{c} ± 2.25	1.39 ^{d} ± 0.01	28.76 ^{c} ± 0.77	2.155 ± 0.06	43.17 ± 1.89	1.64 ^{d} ± 0.09	33.05 ^{c} ± 2.67	1.37 ^{c} ± 0.06	30.44 ^{c} ± 2.19	2.32 ± 0.12	48.27 ± 3.76
	Moringa oleifera	1.36 ^{d} ± 0.03	29.25 ^{c} ± 2.15	1.25 ^{e} ± 0.04	27.28 ^{c} ± 1.78	1.98 ± 0.05	42.89 ± 3.04	1.45 ^{e} ± 0.04	31.05 ^{c} ± 2.06	1.45 ^{b} ± 0.01	28.71 ^{c} ± 1.81	2.17 ± 0.045	45.40 ± 2.08
	Vachellia nilotica	2.11 ^{a} ± 0.01	43.08 ^{a} ± 1.16	1.86 ^{a} ± 0.04	38.95 ^{a} ± 1.48	3.04 ± 0.03	62.55 ± 1.9	2.30 ^{a} ± 0.02	47.07 ^{a} ± 1.75	1.95 ^{a} ± 0.04	40.78 ^{a} ± 1.42	3.27 ± 0.04	67.46 ± 2.46

Means with similar letters are not statiscally different at p 0.05.

4. Discussion

The present study aimed to assess the capability of forest tree species in the land around and inside the industrial sites in Pakistan for regulation and sequestering of carbon to mitigate CO_2 emissions within the surroundings of various industrial locations. In Pakistan, around most industrial sites, many barren lands and conventional agriculture-practicing farms can be invested in as a nature-based solutions for mitigating direct CO_2 emissions. The current study emphasized that the species have an immense capacity to sequester carbon generated from industrial surroundings. The findings also point out their tolerance to higher amounts of CO_2 and their capacity to intercept pollutants, thus improving the urban air quality. A total of three industrial sites were analyzed in this particular research to determine carbon sequestration in the trees and soil present inside the 100 m^2 zone of the industries and the 100 m^2 outside of the industrial sites.

Biomass accumulation and growth depend upon site quality and conditions, type of soil on which trees are planted, age, management practices, and their interaction with belowground components [37,38]. In the present study, the maximum growth and biomass accumulation (above- and belowground) was measured for tree species present inside the selected industrial sites compared with outside. The species inside the industry that depicted the maximum above- and belowground biomass was *M. oleifera* (2.24 and 0.58 Mg ha^{-1}), followed by *Syzygium cumini* (2.17 and 0.57 Mg ha^{-1}), and *Eucalyptus camaldulensis* (12.95 and 3.36 Mg ha^{-1}), whereas outside the industrial area, *Vachellia nilotica* and *D. sissoo* trees displayed higher growth and biomass accumulation. The above- and belowground biomass of different tree species of the present study are comparatively lower than the biomass measured by Yasin et al. [39] for *P. deltoides*, Yasin, et al. [40,41] for *D. sissoo*, *B. ceiba*, *P. deltoides*, and *E. camaldulensis*, Yasin, et al. [39] for *V. nilotica* in Pakistan, Kanime et al. [42] for *D. sissoo* and *P. deltoids*, and Faiz, et al. [43] for a eucalyptus hybrid plantation in the Tarai region of India. However, a similar amount of total biomass accumulation has been documented in previous studies in *E. camaldulensis* [44], *A. nilotica*, and *D. sissoo* [45].

Carbon stock is the absolute amount of carbon present at the time of inventory, while carbon sequestration rate refers to the procedure of removing carbon from the atmosphere and dumping it in a carbon pool [46]. Trees, being perennial vegetation, can store an ample amount of carbon both in above- and belowground parts. In the present study, above- and belowground carbon, along with CO_2 sequestration rate, was estimated inside and outside different industries. The total tree carbon stock and sequestration rate inside the industrial area was higher for *E. camaldulensis* (7.83 Mg ha^{-1} and 28.70 Mg ha^{-1}), while in the outside industries, higher carbon stock and sequestration were measured for *V. nilotica* (2.51 Mg ha^{-1} and 9.19 Mg ha^{-1}). The amount of carbon measured in the present study is comparatively much less as compared with the findings of Arora and Chaudhry [47] for *V. nilotica* + *D. sissoo* (41.44 t ha^{-1}) and Zabek and Prescott [48] for *P. deltoids* (51.2 t ha^{-1}) planted in different sub-continent regions. However, our findings are in agreement with the results of Yasin et al. [41], who reported total carbon stock (7.17 t ha^{-1}) in 8-year-old *V. nilotica* trees. Similarly, the CO_2 sequestration is slightly higher than that estimated by Kaul et al. [49] for *P. deltoids* (8 t ha^{-1} yr^{-1}) and Lal and Singh [50] for plantations (3.2 t ha^{-1} yr^{-1}). The carbon sequestering potential of forest tree species in the current study is relevant to research conducted in the USA that depicted the restored lands with forest species around industrial sites as capable of mitigating the emissions generated by them [24].

In a terrestrial ecosystem, the soil is considered an important carbon pool to mitigate atmospheric CO_2. Soil carbon stock depends on the land use pattern, soil type, topography, climatic conditions, and management practices [41,51]. Greater soil carbon is estimated in soils with trees due to more leaf litter [46]. The findings of our study indicated that the amount of soil carbon decreased with the increase in depth: higher soil carbon stock was estimated at 0–20 cm depth both inside and outside of the industries. The fact behind this greater amount of soil carbon in surface soil is due to the higher accumulation of tree litter, which ultimately results in greater carbon input [44,51,52]. Our findings are very consistent with the findings of Arora and Chaudhry [45] for *V. nilotica* and *D. sissoo* and Arora et al. (2014) for *P. deltoids* trees. Overall, inside and outside the industries, maximum SOC was: 53.05 Mg ha^{-1}, 66.16 Mg ha^{-1} (0–20 cm), 43.11 Mg ha^{-1}, and 47.51 Mg ha^{-1} (20–40 cm) for *D. sissoo*. The above values are comparatively higher than those reported by Yasin et al. [41] in 8-year-old *V. nilotica* trees (26.27 Mg ha^{-1}) at 0–15 cm depth. However, our findings are very consistent with the findings of Yasin et al. [39], who reported soil carbon stock of 38.57 Mg ha^{-1}, 41.81 Mg ha^{-1}, and 43.73 Mg ha^{-1} at 0–30 cm depth in 6-year-old *P. deltoids* trees across three tehsils of the Chiniot district.

5. Policy Implications and Conclusions

Increasing global warming induced by industries is an eminent source of increasing disasters and the spread of new diseases. The developing world has put very little effort into providing their industries with clean and green emissions. The most common, cheap, and less laborious strategy of mitigating industrial emissions are nature-based solutions. In support of various previous research, the current study has proven nature-based solutions to mitigate carbon emissions. The current findings depict that tree species present inside and outside the vicinity of various industries have strong potential in mitigating air emissions.

The national government is actively participating in forming industrial and economic zones in the country that could, later on, increase pollution levels. The current study recommends that plantation of native and fast-growing trees in and around these zones be made mandatory. This study indicates a huge potential of natural-based solutions on a national level to mitigate the direct air pollution emissions from the industrial sites in the suburbs of big cities and make them locally sustainable. Furthermore, it recommends the government and industrial sector rely on nature-based green solutions and employ greener technology to ensure amplified mitigation of carbon emissions.

The current study acts as a starting point for the industrial sector to explore the viability and practicability of applying nature and technology-based solutions to mitigate emissions. Additional and more focused work on this aspect can lead the industrial community in the region to use these studies as a foundation for identifying facilities in and around the vicinity of industries for the growth of vegetation for mitigating emissions.

Author Contributions: M.Z., G.Y. and A.J.: Conceptualization, Formal analysis, Writing—Original Draft Preparation. S.K.Q., A.U.H. and M.Y.: Resources, data curation, Field work, Methodology. S.U.R. and W.G.: Writing—Review and Editing, Visualization, Validation and Funding Acquisition. All authors have read and agreed to the published version of the manuscript.

Funding: The National Natural Science Foundation of China (No. 51709265).

Institutional Review Board Statement: The study did not require ethical approval, so chose to exclude this statement.

Informed Consent Statement: Not applicable.

Data Availability Statement: All data sets are associated with the paper. Moreover, an excel sheet for raw data will be provided on request after acceptance of the paper.

Conflicts of Interest: On behalf of the first author and remaining co-authors, I declare that no one has any conflict of interest.

References

1. El Zein, A.L.; Chehayeb, N.A. The effect of greenhouse gases on earth's temperature. *Int. J. Environ. Monit. Anal.* **2015**, *3*, 74. [CrossRef]
2. Białecki, R.; Stanek, W. Cumulative Green-House Gasses (GHG) Emissions as Total Measure of Global Warming Potential. In *Thermodynamics for Sustainable Management of Natural Resources*; Springer: Berlin, Germany, 2017; pp. 473–488.
3. Zhou, T.; Roorda, M.J.; MacLean, H.L.; Luk, J. Life cycle GHG emissions and lifetime costs of medium-duty diesel and battery electric trucks in Toronto, Canada. *Transp. Res. Part D Transp. Environ.* **2017**, *55*, 91–98. [CrossRef]

4. Sinha, R.K.; Chaturvedi, N.D. A review on carbon emission reduction in industries and planning emission limits. *Renew. Sustain. Energy Rev.* **2019**, *114*, 109304. [CrossRef]
5. Mohajerani, A.; Bakaric, J.; Jeffrey-Bailey, T. The urban heat island effect, its causes, and mitigation, with reference to the thermal properties of asphalt concrete. *J. Environ. Manag.* **2017**, *197*, 522–538. [CrossRef] [PubMed]
6. Vidale, S.; Campana, C. Ambient air pollution and cardiovascular diseases: From bench to bedside. *Eur. J. Prev. Cardiol.* **2018**, *25*, 818–825. [CrossRef] [PubMed]
7. Raza, A.; Dahlquist, M.; Jonsson, M.; Hollenberg, J.; Svensson, L.; Lind, T.; Ljungman, P.L. Ozone and cardiac arrest: The role of previous hospitalizations. *Environ. Pollut.* **2019**, *245*, 1–8. [CrossRef]
8. Tabinda, A.B.; Anjum, Z.; Yasar, A.; Rasheed, R.; Mahmood, A.; Iqbal, A. Determination and dispersion of pollutants from different fuel types used in brick kilns by using Gaussian's plume model. *Energy Sources Part A Recovery Util. Environ. Eff.* **2019**, *41*, 1022–1028. [CrossRef]
9. Shakir, S.K.; Azizullah, A.; Murad, W.; Daud, M.K.; Nabeela, F.; Rahman, H.; ur Rehman, S.; Häder, D.-P. Toxic metal pollution in Pakistan and its possible risks to public health. *Rev. Environ. Contam. Toxicol.* **2016**, *242*, 1–60.
10. Alam, K.; Mukhtar, A.; Shahid, I.; Blaschke, T.; Majid, H.; Rahman, S.; Khan, R.; Rahman, N. Source apportionment and characterization of particulate matter (PM10) in urban environment of Lahore. *Aerosol Air Qual. Res.* **2014**, *14*, 1851–1861. [CrossRef]
11. Shams, T.; Khwaja, M.A. *Assessment of Pakistan National Ambient Air Quality Standards (NAAQS's) with Selected Asian Countries and WHO*; Sustainable Development Policy Institute: Islamabad, Pakistan, 2019.
12. WHO. *Estimated Deaths & DALYs Attributable to Selected Environmental Risk Factors, by WHO Member State, 2002*; Department of Public Health & Environnment: Denver, CO, USA, 2007.
13. Kondo, R.; Kinoshita, Y.; Yamada, T. Green procurement decisions with carbon leakage by global suppliers and order quantities under different carbon tax. *Sustainability* **2019**, *11*, 3710. [CrossRef]
14. Lau, W.W. Beyond carbon: Conceptualizing payments for ecosystem services in blue forests on carbon and other marine and coastal ecosystem services. *Ocean. Coast. Manag.* **2013**, *83*, 5–14. [CrossRef]
15. Cohen-Shacham, E.; Walters, G.; Janzen, C.; Maginnis, S. *Nature-Based Solutions to Address Global Societal Challenges*; IUCN: Gland, Switzerland, 2016; Volume 97, pp. 2016–2036.
16. Abson, D.J.; Von Wehrden, H.; Baumgärtner, S.; Fischer, J.; Hanspach, J.; Härdtle, W.; Heinrichs, H.; Klein, A.; Lang, D.; Martens, P. Ecosystem services as a boundary object for sustainability. *Ecol. Econ.* **2014**, *103*, 29–37. [CrossRef]
17. Bridgewater, P. Whose nature? What solutions? Linking Ecohydrology to Nature-based solutions. *Ecohydrol. Hydrobiol.* **2018**, *18*, 311–316. [CrossRef]
18. Kabisch, N.; Frantzeskaki, N.; Pauleit, S.; Naumann, S.; Davis, M.; Artmann, M.; Haase, D.; Knapp, S.; Korn, H.; Stadler, J. Nature-based solutions to climate change mitigation and adaptation in urban areas: Perspectives on indicators, knowledge gaps, barriers, and opportunities for action. *Ecol. Soc.* **2016**, *21*, 15. [CrossRef]
19. Hewitt, C.N.; Ashworth, K.; MacKenzie, A.R. Using green infrastructure to improve urban air quality (GI4AQ). *Ambio* **2020**, *49*, 62–73. [CrossRef] [PubMed]
20. Carstens, D.; Amer, R. Spatio-temporal analysis of urban changes and surface water quality. *J. Hydrol.* **2019**, *569*, 720–734. [CrossRef]
21. Sæbø, A.; Janhäll, S.; Gawronski, S.W.; Hanslin, H.M. *Urban Forestry and Pollution Mitigation*; Routledge: London, UK, 2017.
22. Bealey, W.; McDonald, A.; Nemitz, E.; Donovan, R.; Dragosits, U.; Duffy, T.; Fowler, D. Estimating the reduction of urban PM10 concentrations by trees within an environmental information system for planners. *J. Environ. Manag.* **2007**, *85*, 44–58. [CrossRef] [PubMed]
23. Escobedo, F.J.; Nowak, D.J. Spatial heterogeneity and air pollution removal by an urban forest. *Landsc. Urban Plan.* **2009**, *90*, 102–110. [CrossRef]
24. Nowak, D.J. Institutionalizing urban forestry as a "biotechnology" to improve environmental quality. *Urban For. Urban Green.* **2006**, *5*, 93–100. [CrossRef]
25. Gopalakrishnan, V.; Ziv, G.; Hirabayashi, S.; Bakshi, B.R. Nature-based solutions can compete with technology for mitigating air emissions across the United States. *Environ. Sci. Technol.* **2019**, *53*, 13228–13237. [CrossRef]
26. Kroeger, T.; Escobedo, F.J.; Hernandez, J.L.; Varela, S.; Delphin, S.; Fisher, J.R.; Waldron, J. Reforestation as a novel abatement and compliance measure for ground-level ozone. *Proc. Natl. Acad. Sci. USA* **2014**, *111*, E4204–E4213. [CrossRef] [PubMed]
27. Alonso, R.; Vivanco, M.G.; González-Fernández, I.; Bermejo, V.; Palomino, I.; Garrido, J.L.; Elvira, S.; Salvador, P.; Artíñano, B. Modelling the influence of peri-urban trees in the air quality of Madrid region (Spain). *Environ. Pollut.* **2011**, *159*, 2138–2147. [CrossRef]
28. McHale, M.R.; McPherson, E.G.; Burke, I.C. The potential of urban tree plantings to be cost effective in carbon credit markets. *Urban For. Urban Green.* **2007**, *6*, 49–60. [CrossRef]
29. Akbar, K.; Ashraf, I.; Shakoor, S. Analysis of urban forest structure, distribution and amenity value: A case study. *J. Anim. Plant Sci.* **2014**, *24*, 1636–1642.
30. Cairns, M.A.; Brown, S.; Helmer, E.H.; Baumgardner, G.A. Root biomass allocation in the world's upland forests. *Oecologia* **1997**, *111*, 1–11. [CrossRef] [PubMed]
31. Thomas, S.C.; Martin, A.R. Carbon content of tree tissues: A synthesis. *Forests* **2012**, *3*, 332–352. [CrossRef]

32. Afzal, M.; Aqeela, M.A. Factors affecting carbon sequestration in trees. *J. Agric. Res.* **2013**, *51*, 6168.
33. Ravindranath, N.H.; Ostwald, M. *Carbon Inventory Methods: Handbook for Greenhouse Gas Inventory, Carbon Mitigation and Roundwood Production Projects*; Springer Science & Business Media: Berlin/Heidelberg, Germany, 2007; Volume 29.
34. Brown, S.; Gillespie, A.J.; Lugo, A.E. Biomass estimation methods for tropical forests with applications to forest inventory data. *For. Sci.* **1989**, *35*, 881–902.
35. Rai, S. Above ground biomass in tropical rain forests of Western Ghats, India. *Indian For.* **1984**, *110*, 754–764.
36. Hawkins, T. *Biomass and Volume Tables for Eucalyptus camaldulensis, Dalbergia sissoo, Acacia auriculiformis and Cassia siamea in the Central Bhabar-Terai of Nepal*; University of Oxford: Oxford, UK, 1987.
37. de M. Sá, J.C.; Cerri, C.C.; Dick, W.A.; Lal, R.; Filho, S.P.V.; Piccolo, M.C.; Feigl, B.E. Organic matter dynamics and carbon sequestration rates for a tillage chronosequence in a Brazilian Oxisol. *Soil Sci. Soc. Am. J.* **2001**, *65*, 1486–1499.
38. JHA, K.K. Root Carbon Sequestration and Its Efficacy in Forestry and Agroforestry Systems: A Case of Populus euramericana I-214 Cultivated in Mediterranean condition. *Not. Sci. Biol.* **2018**, *10*, 68–78. [CrossRef]
39. Yasin, G.; Nawaz, M.; Siddiqui, M.; Niazi, N. Biomass, carbon stocks and CO_2 sequestration in three different aged irrigated populus deltoides bartr. Ex marsh. Bund planting agroforestry systems. *Appl. Ecol. Environ. Res.* **2018**, *16*, 6239–6252. [CrossRef]
40. Yasin, G.; Nawaz, M.F.; Martin, T.A.; Niazi, N.K.; Gul, S.; Yousaf, M.T.B. Evaluation of Agroforestry Carbon Storage Status and Potential in Irrigated Plains of Pakistan. *Forests* **2019**, *10*, 640. [CrossRef]
41. Yasin, G.; Nawaz, M.F.; Yousaf, M.T.B.; Gul, S.; Qadir, I.; Niazi, N.K.; Sabir, M.A. Carbon stock and CO_2 sequestration rate in linearly planted Vachellia nilotica farm trees. *Pak. J. Agric. Sci.* **2020**, *57*, 807–814.
42. Kanime, N.; Kaushal, R.; Tewari, S.; Raverkar, K.; Chaturvedi, S.; Chaturvedi, O. Biomass production and carbon sequestration in different tree-based systems of Central Himalayan Tarai region. *For. Trees Livelihoods* **2013**, *22*, 38–50. [CrossRef]
43. Faiz, M.; Singh, R.; Jattan, S.; Singh, K. Root studies in a Eucalyptus hybrid plantation at various ages. *Indian For.* **2000**, *126*, 1165–1174.
44. Nawaz, M.F.; Mazhar, K.; Gul, S.; Ahmad, I.; Yasin, G.; Asif, M.; Tanvir, M. Comparing the early stage carbon sequestration rates and effects on soil physico-chemical properties after two years of planting agroforestry trees. *J. Basic Appl. Sci.* **2017**, *13*, 527–533. [CrossRef]
45. Kaur, B.; Gupta, S.; Singh, G. Carbon storage and nitrogen cycling in silvopastoral systems on a sodic in northwestern India. *Agrofor. Syst.* **2002**, *54*, 21–29. [CrossRef]
46. Takimoto, A.; Nair, P.R.; Nair, V.D. Carbon stock and sequestration potential of traditional and improved agroforestry systems in the West African Sahel. *Agric. Ecosyst. Environ.* **2008**, *125*, 159–166. [CrossRef]
47. Arora, P.; Chaudhry, S. Vegetation and soil carbon pools of mixed plantation of Acacia nilotica and Dalbergia sissoo under social forestry scheme in Kurukshetra, India. *J. Mater. Environ. Sci.* **2017**, *8*, 4565–4572. [CrossRef]
48. Zabek, L.; Prescott, C. Biomass equations and carbon content of above-ground leafless biomass of hybrid poplar in Coastal British Columbia. *For. Ecol. Manag.* **2006**, *223*, 291–302. [CrossRef]
49. Kaul, M.; Mohren, G.; Dadhwal, V. Carbon storage and sequestration potential of selected tree species in India. *Mitig. Adapt. Strateg. Glob. Change* **2010**, *15*, 489–510. [CrossRef]
50. Lal, M.; Singh, R. Carbon sequestration potential of Indian forests. *Environ. Monit. Assess.* **2000**, *60*, 315–327. [CrossRef]
51. Arora, G.; Chaturvedi, S.; Kaushal, R.; Nain, A.; Tewari, S.; Alam, N.M.; Chaturvedi, O.P. Growth, biomass, carbon stocks, and sequestration in an age series of Populus deltoides plantations in Tarai region of central Himalaya. *Turk. J. Agric. For.* **2014**, *38*, 550–560. [CrossRef]
52. Nawaz, M.; Yousaf, M.; Yasin, G.; Gul, S.; Ahmed, I.; Abdullah, M.; Rafay, M.; Tanvir, M.; Asif, M.; Afzal, S. Agroforestry status and its role to sequester atmospheric CO_2 under semi-arid climatic conditions in Pakistan. *Appl. Ecol. Environ. Res.* **2018**, *16*, 645–661. [CrossRef]

land

MDPI

Article

Social Relevance of Ecosystem Services Provided by Urban Green Infrastructures: A Mixed Qualitative–Quantitative Case Study Approach

Sofia Baldessari [1], Isabella De Meo [2], Maria Giulia Cantiani [3,*] and Alessandro Paletto [1]

1 Consiglio per la ricerca in agricoltura e l'analisi dell'economia agraria (CREA), Research Centre for Forestry and Wood, 38123 Trento, Italy; sofia.baldessari@crea.gov.it (S.B.); alessandro.paletto@crea.gov.it (A.P.)
2 Consiglio per la ricerca in agricoltura e l'analisi dell'economia agraria (CREA), Research Centre for Agriculture and Environment, 50023 Firenze, Italy; isabella.demeo@crea.gov.it
3 Department of Civil, Environmental and Mechanical Engineering, University of Trento, 38123 Trento, Italy
* Correspondence: maria.cantiani@unitn.it

Abstract: Urban green infrastructures (UGIs) are important elements of the urban matrix providing ecological functions and several ecosystem services beneficial to citizens. Recently, their contribution to the well-being and quality of life of citizens has been widely recognized by both the scientific community and policymakers. This study aims to explore the social relevance of UGI ecosystem services through a mixed qualitative–quantitative case study approach. First, a quantitative literature review was conducted using bibliometric network analysis, followed by a participatory process through a Scientific Café with a sample of Italian stakeholders involved in the UGIs' planning and management. The bibliometric network analysis identified 443 documents (from 2000 to 2024) primarily focused on three ecosystem services provided by UGIs—climate change mitigation, biodiversity conservation, and cultural services, such as health benefits, aesthetic, and recreation. The qualitative results of Scientific Café highlight a particular interest of stakeholders in cultural services such as physical and mental benefits provided by UGIs, as well as improving the urban aesthetic value. Additionally, the role of UGIs in microclimate regulation and heat mitigation was emphasized. The combined analysis of perspectives from the scientific community, policy-makers, stakeholders, and citizens provides a foundation for developing a participatory governance approach to UGIs.

Keywords: bibliometric network analysis; Scientific Café; participatory process; focus group discussions; green areas; Italy

1. Introduction

In recent decades, the key role of green infrastructures in improving the quality of life and well-being of urban populations has become increasingly interesting in the eyes of both the scientific community and policymakers worldwide [1,2]. In 2020, the World Health Organization (WHO) recognized the contribution of green spaces to maintaining citizens' mental and physical well-being [3], while the 2030 Agenda for Sustainable Development emphasized the importance of ensuring universal access to safe, inclusive and accessible green and public spaces by 2030. In accordance with the Goal 11, Target 11.7 of the Sustainable Development Goals (SDGs), the inclusiveness and accessibility of green areas must be ensured especially for women and children, older persons, and persons with disabilities [4,5]. At the European level, the environmental policy of the European Union (EU) has promoted green infrastructure both for rural and urban areas within its EU

 https://doi.org/10.3390/land14020230

Biodiversity Strategy for 2030, emphasizing the pivotal role in supporting Trans-European network connectivity and enhancing the provision of ecosystem services [6]. Furthermore, in 2021 the European Commission promoted the Green City Accord (GCA)—a movement of mayors of European large cities—to improve the quality of life for all Europeans and accelerate the implementation of the European Green Deal (2020). Among the priorities of GCA, there is the conservation and enhancement of urban biodiversity, which can be achieved by increasing the extension and quality of green areas in cities [7,8].

Green infrastructures have been defined by the European Commission (2014) as a strategically planned network of high-quality natural and semi-natural areas with other environmental features, designed and managed to deliver a wide range of ecosystem services and protect biodiversity in both rural and urban settings [9]. The role of green infrastructures is of pivotal importance in urban areas due to the global phenomenon of population growth and urbanization [10]. In fact, the world's population is expected to increase to 9.8 billion by 2050, and the world's population living in urban areas will reach 68% of the total by 2050 [11]. In Europe, currently urbanized areas consisting of human settlements with a high population density cover 22% of EU territory; this is the result of a growth trend over the last 20 years that has led to a 3.4% increase in urbanized areas per decade [7]. Regarding the population, an increase of 30 million additional people in EU countries is expected by 2050 [12]. In a context of increasing urbanization and population density, urban green infrastructures (UGIs)—e.g., parks, gardens, allotments, community gardens, cemeteries, green roofs, urban orchards, urban and peri-urban forests—are assuming an ever-increasing importance [13].

In accordance with the definition provided by the European Commission [9], the UGIs provide several ecosystem services (ESs) essential for the psychological and physical well-being of the urban population, such as heat mitigation, noise reduction, flood protection, rainwater runoff regulation, microclimate and air quality regulation, outdoor recreation and aesthetic pleasure [14,15]. All these ESs fall into the four categories described by the Millennium Ecosystem Assessment (MEA) in 2005, as follows [16]: provisioning services (the products obtained from ecosystems for basic human needs); regulating services (those that control the states and rates of physical and biotic systems and processes in ways that are beneficial to humans); cultural services (the nonmaterial benefits that humans obtain from ecosystems); and supporting services (those necessary for the production of all other ecosystem services).

In the literature, many studies explored the importance of UGIs in the provision of different ESs from the biophysical point of view [17]. Some studies have focused on the water storage and water retention capacity of UGIs and the reduction in air temperature through evapotranspiration [18,19], while others have considered the role of UGIs in reducing the heat island and pollution island effects [20,21]. Furthermore, a more limited number of studies investigated the monetary value of environmental benefits provided by UGIs [22,23]. However, only a few studies have considered as their main focus the social perception of ESs provided by UGIs. Among these studies, Giannico et al. analyzed citizens' perceptions of the benefits provided by UGIs in 51 European cities [24], while Ostoić et al. focused on citizens' perceptions of the current state of urban forests and green spaces in seven Southeast European cities (i.e., Zagreb, Novi Sad, Beograd, Banja Luka, Sarajevo, Podgorica, Skopje) [25]. Recently, Molari et al. investigated citizens' perceptions of green walls in terms of attractiveness, their integrative role in the environment, and comfortable space in Turin and Lisbon [26]. In the Asian context, Kim et al. conducted a study on residents' perceptions of informal green spaces in Ichikawa city (Japan) [27], while Chen et al. analyzed the ESs (i.e., air pollution removal, temperature reduction, rainwater runoff regulation, noise abatement) demand related to the UGIs in Guangzhou in southern China [28].

Starting from these considerations, the aim of the present study was to investigate the social relevance of ESs provided by UGIs through a mixed qualitative–quantitative case study approach. Firstly, a literature review was carried out, and subsequently a participatory process was undertaken in a case study in Italy. The main research question was whether the relevance of ESs assigned by the international scientific community (through literature review) converges with or differs from that assigned by local stakeholders (through an Italian participatory case study process). The present research was developed in the context of the Horizon Europe Project ForestValue2. ForestValue2 brings together owners and managers of national and regional Research, Development and Innovation (RDI) programs in eleven Member States of the European Union (EU) and in one Associated Country, with the aim of contributing to the alignment of national research and innovation policies.

2. Materials and Methods

The present study was structured in two steps. In the first step, a Bibliometric Network Analysis (BNA) was performed to investigate ESs provided by UGIs according to the international literature and producing quantitative data. The second step consisted of a qualitative analysis of ESs based on the opinions provided by a sample of Italian stakeholders involved through a participatory process and a related event (Scientific Café). The main ESs provided by UGIs that emerged from the BNA and the Scientific Café were compared to highlight convergences and divergences between the international scientific community and the stakeholders in an Italian case study.

A mixed qualitative–quantitative approach was adopted in this study in order to provide a more comprehensive understanding of a complex phenomenon, such as the role of UGIs in the provision of ESs useful for the well-being and quality of life of citizens. The quantitative data provided by the literature can offer a broad picture of the phenomenon of ESs provided by UGIs, while the qualitative information provided by stakeholders can offer an in-depth analysis of interventions and actions to improve the delivery of individual ESs. This type of data allows for the exploration of individuals' motivations, personal experiences and perceptions, which are often beyond the scope of quantitative methods.

The integration and comparison of the main ESs analyzed by scientific literature (quantitative approach based on bibliometric network analysis) with those perceived as more important by a sample of Italian stakeholders (qualitative approach) is the innovative aspect of this study.

2.1. Bibliometric Network Analysis: Methodology

The BNA is based on the combination of the bibliometric approach with the social network approach; the first one is used to analyze scientific productivity on a specific topic, while the second one is aimed at understanding the relationships among all components (e.g., keywords, concepts) of a system to identify the key role of some components of the system itself [29,30].

In the present study, the BNA aimed to identify and examine the international literature on ESs provided by UGIs. The peer-review publications were retrieved from the Scopus database (https://www.scopus.com, accessed on 2 September 2024) using the following string of keywords: ("urban forest*" OR "peri-urban forest*" OR "urban greening*" OR "urban green infrastructure*") AND ("perception*" OR "preference*" OR "social") AND ("ecosystem service*" OR "environmental service*"). These keywords were searched in the titles, abstracts, and keywords of individual publications, considering as a timeframe the last 25 years (period from 2000 to 2024). All data were exported as "comma-separated values" (CSV) files and processed using the VOSviewer software (version 1.6.17). The VOSviewer is an open-access software used to create, visualize, and explore maps combining bibliometric reviews and network analysis [31]. The main output provided by VOSviewer is the network of connections among the bibliometric data grouped in clusters.

In this study, co-occurrence analyses were conducted to create network maps of the keywords used in the literature, highlighting the main ESs provided by green areas in the urban context investigated by the international literature. The results provided by the co-occurrence analysis can help us to identify the most important and recurring keywords. Consequently, the concepts investigated by the international scientific community are obtained and evidenced. The BNA on ESs provided by the UGIs was implemented using the two following indicators [31]:

- Co-occurrence (O)—the number of co-occurrences of two keywords is the number of publications in which both keywords occur together in the title, abstract, or keyword list;
- Total link strength (TLS)—the cumulative strength of the links of an item with other items.

The values of co-occurrence and total link strength of the single ecosystem service analyzed in the literature were used as a proxy for their importance in the eyes of the scientific community.

2.2. *Scientific Café: Methodology*

In the second step of the study, based on a qualitative case study approach, a participatory process was conducted to analyze the perception of a sample of Italian stakeholders on the relevance of ESs provided by green areas located in the urban context. Among the different methodologies that can be used in the participatory process, based on the utilization of diverse techniques, in the present research, a Scientific Café—also known as World Cafe or Apéritifs Scientifique—was conducted. As stated by Nesseth et al. [32], Scientific Café is a public discussion of socially pertinent questions and needs that have scientific content in an informal setting, with instruments to ensure effective and well-structured communication between scientists, practitioners, civil society, and decision-makers. In other words, these events are informal occasions for the general public (citizens) and/or stakeholders to meet scientists—e.g., researchers and experts—and to discuss scientific topics of interest to them [33]. According to Baldessari et al. [34], Scientific Café encourages informal, open discussions in small groups (less than 50 participants), with the advantage of bringing out new ideas and initiatives as a result of joint discussion.

In this study, a Scientific Café was conducted on 11 September 2024 in Italy in the city of Padua (Northeast Italy). Following the rule of organizing these forums external to the academic environment to favor an informal environment and facilitate interaction, the Café was conducted within a National Congress of Forests. This contributed to the success of the event, given high number of participants who were attending the Congress who could also participate to the Scientific Café.

The event was organized from June to August 2024, considering the following steps proposed by De Meo et al. [35]: (i) identification of the scientific topic; (ii) selection of the audience; (iii) logistical organization of the event; (iv) definition of the participatory technique; and (v) definition of the desired outcomes.

Concerning the topic, as previously mentioned, the scientific issue discussed was "The ESs provided by UGIs", and in particular, how they are perceived by citizens and the possible management actions that can be taken to valorize their role in the UGIs.

The selected audience were stakeholders with a direct or indirect interest in the planning, management, and governance of UGIs, which were directly invited to participate by the research team or informed of the Scientific Café by reading the Congress program. In particular, technicians of public green areas, freelancers who deal with green spaces design, and representatives of private companies that manage green spaces were invited to participate in the event. The invitations to stakeholders were made following a stakeholder analysis—conducted between 20 and 31 August 2024—that allowed us to identify all the key stakeholders of the region (Veneto) and of the three surrounding regions (Friuli-Venezia Giulia, Trentino-Alto Adige, Emilia-Romagna). The stakeholder analysis, applied following the methodology proposed by Grilli et al. [36], is a technique that is used to identify all groups of people, organized or unorganized, who share a stake in a particular issue [37].

Regarding the logistical organization of the event, an in-person Scientific Café was preferred to collect a greater amount of information from the participants. Then, the organizing group established the schedule, including the overall duration of the Scientific Café, the length of expert interventions and the time allocated for the participatory process moderated by the facilitator. An expert on the topic and a facilitator were involved to support the organizing staff in introducing the issue and facilitating the event. Afterwards, the "Problem Tree" and "Strategy Tree" techniques were employed [35]. The participants, working individually or in small groups, wrote down on notes what they considered to be the main critical issues concerning the ESs provided by UGIs, which were then grouped with the facilitators by theme; ideas for solutions and strategies to overcome the elements of weakness were also proposed. At the end of the session, the facilitators presented a summary of the concepts that emerged.

The key ESs related to UGIs that need to be valorized through planning and management and the key actions undertaken to pursue such valorization are the main outputs of the Scientific Café. The Scientific Café participants answered based on their knowledge and skills; therefore, the outputs are mainly referred to the Italian context.

3. Results

3.1. Bibliometric Network Analysis: Results

The results of the BNA show 443 peer-review publications on ESs provided by UGIs on the Scopus database. The first publications on this topic date back to 1999, focusing on the role of agroforestry systems and tree forests in offering environmental benefits to local communities [38]. However, the topic started to attract the attention of the scientific community in 2010, from which year an increasing number of publications per year has been noted (Figure 1). In particular, an average number of publications of 40.2 per year (SD = 17.3) is recorded for the period 2015–2024, with a range between a minimum of 17 publications in 2017 and a maximum of 60 in 2021.

Observing the data on the relevance of ESs by category, the results of the BNA show sixteen keywords related to cultural services, nine related to regulating services, six related to supporting services, and only three related to provisioning services (Table 1). However, the ecosystem service with the highest TLS value is climate change mitigation in the regulating services category (O = 26; TLS = 309), followed by biodiversity in the supporting services category (O = 29; TLS = 296) and cultural ecosystem services (O = 22; TLS = 198). When aggregating keywords by category of ES, the results highlight that academics and researchers have found three main services in the category of cultural services: landscape aesthetics; outdoor recreation; physical and psychological well-being of citizens related to UGIs. Concerning regulating services, the scientific community has mainly focused on climate change mitigation. In fact, except for water management, all the keywords refer to climate change, air quality and related issues. Observing the category of supporting services, almost all publications have considered the various aspects of biodiversity. In addition, it is interesting to underline the low importance given to the provisioning services provided by UGIs. In fact, the few studies found have considered only three ESs—food production (O = 6; TLS = 65), the supply of woody biomass for energy use (O = 4; TLS = 61), and timber (O = 5; TLS = 57).

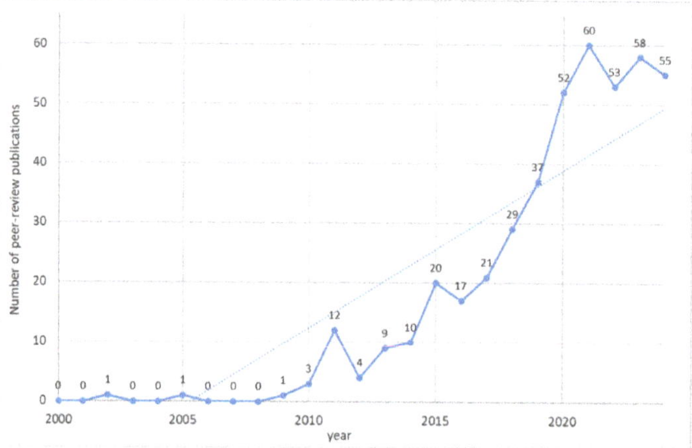

Figure 1. Trend of peer-reviewed publications on ESS provided by UGIs from 2000 to 2024.

Table 1. Main keywords directly related to ESs provided by UGIs in the international literature.

Ecosystem Service	Co-Occurrence (O)	Total Link Strength (TLS)
	Cultural services	
Cultural ecosystem services (plural)	22	198
Aesthetics	9	124
Cultural ecosystem service (singular)	7	118
Public health	9	118
Quality of life	10	113
Landscape	8	99
Well-being	6	89
Recreation	9	85
Recreational activity	6	76
Culture	4	65
Psychology	5	64
Landscape structure	5	59
Outdoor recreation	4	46
Landscape change	6	40
Well-being	4	40
Cultural services	4	17
	Regulating services	
Climate change	26	309
Carbon sequestration	11	158
Air quality	7	114
Carbon	4	86
Water management	7	93
Heat island	7	81
Urban heat island	6	56
Microclimate	4	55
Carbon storage	4	51

Table 1. *Cont.*

Ecosystem Service	Co-Occurrence (O)	Total Link Strength (TLS)
	Supporting services	
Biodiversity	29	296
Urban biodiversity	10	61
Biodiversity conservation	5	59
Connectivity	6	59
Species diversity	5	52
Species richness	4	16
	Provisioning services	
Food production	6	65
Biomass	4	61
Timber	5	57

Figure 2 shows co-occurrence network maps with only the keywords directly related to ESs. Considering a minimum number of occurrences of a keyword equal to 5, of the 2888 total keywords identified, 221 meet the threshold. In the first step, the 221 keywords were divided into six clusters, and, subsequently, in the second step, 40 keywords directly related to the urban ESs belonging to five clusters were identified (Figure 2). With reference to this last network, the blue cluster includes most of the studies on cultural services, such as landscape aesthetic, recreation, well-being and health, while the purple cluster is exclusively focused on the mitigation of heat islands and the related quality of life of the residents. The other three clusters include those studies that have simultaneously considered multiple ESs or the synergies/trade-offs between them: carbon storage/sequestration and biomass use for energy in the red cluster; biodiversity and climate change mitigation in the green cluster; and outdoor recreation and biodiversity conservation in the yellow cluster.

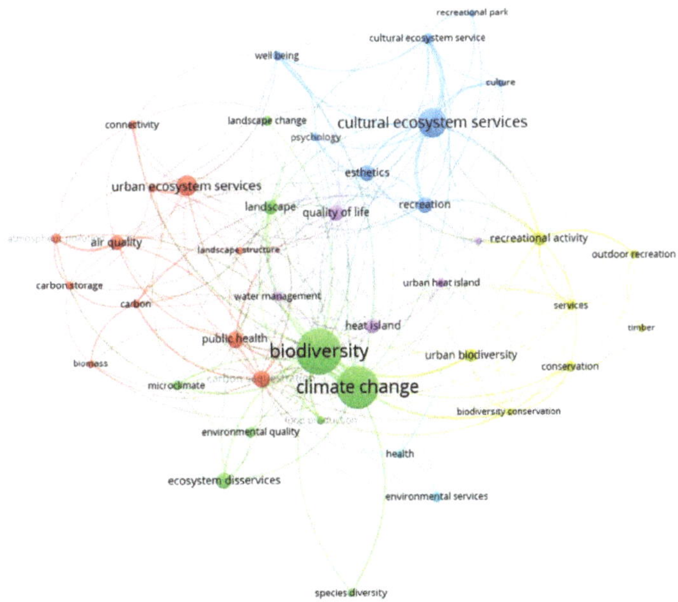

Figure 2. Co-occurrence network map considering the keywords directly related to the ESs provided by UGIs.

3.2. Scientific Café: Results

At the Scientific Café, 40 stakeholders, including technicians of public administrations, freelancers working in urban green design studios, and private sector operators who deal with the management and maintenance of urban greenery, attended the event. Regarding the background of the participants, approximately 75% of the participants have a background in agricultural and forestry sciences/engineering, followed by 15% in environmental/natural sciences, and 5% in architecture/urban planning and social and economic sciences, respectively. The Scientific Café lasted a total of two and a half hours, with a 15 min break in between.

The Scientific Café produced as its main result the list of ESs provided by UGIs that should be valorized through management plans and specific actions (see Table 2).

Table 2. The ESs provided by UGIs and key actions for their valorization in accordance with the stakeholders' opinions.

Ecosystem Services	Key Actions
Regulating services	
Heat islands mitigation	- Creating large-scale UGIs in urban areas
Climate change adaptation	- Selection of trees and shrubs in UGIs tolerant to climate change
Cultural services	
Public health, well-being and quality of life	- Choice of allergy-free trees and shrubs in UGIs
Aesthetic value	- Diversification of tree species composition, structure, and layers (grass, shrub, tree) - Restoring abandoned infrastructure areas (industrial buildings, ruins) through green spaces
Outdoor recreation	- Co-design UGIs with citizens to improve aesthetic value and recreational opportunities according to users' preferences - Involving citizens in the maintenance of green areas, promoting greater participation and social cohesion
Supporting services	
Urban biodiversity (flora and fauna)	- Foster ecological connectivity by creating larger UGIs (e.g., urban forests) and building green corridors - Selection of native tree and shrub species in order to favor more mixed UGIs - Selection of fruticose tree and shrub species as a food source for wildlife - Maintaining tree-related microhabitats if they are not dangerous to public safety

According to the sample of stakeholders, the key ESs that can be enhanced through the sustainable and rational management of UGIs are the improvement of air quality and the mitigation of heat islands. In particular, the participants at the Scientific Café did not explicitly mention the roles of UGIs in carbon storage and carbon dioxide (CO_2) sequestration, but repeatedly underlined the roles of parks, gardens, street trees, and urban and peri-urban forests in improving air quality. In accordance with the stakeholders' perspective, the key actions taken to improve green areas for microclimate regulation and air quality depend on the choices of tree, shrub and grass species associated with growth conditions. The sizes of urban parks and forests in the cities used for microclimate and heat mitigation were highlighted as amongst the key aspects by the participants.

Another aspect that emerged during the Scientific Café is the role of UGIs in the health, well-being and quality of life of residents. One keyword mentioned by some participants is "public health", stressing that the importance of green areas in an urban context has grown in the eyes of citizens, especially following the restrictions imposed by the COVID-19 pandemic.

According to the stakeholders, UGIs are currently considered by citizens as a place to relax and reduce daily stress and anxiety in a natural environment close to home.

Based on the discussion during the Café, it is a widespread opinion among stakeholders that UGIs are valued more highly by citizens for their aesthetic value than for other key ecosystem functions. Therefore, to satisfy citizens' requests, the aesthetic value of UGIs should be taken into careful consideration during both the design and daily management stages.

The role of UGIs in providing recreational and leisure opportunities was mentioned by participants in relation to physical well-being and social inclusion. Regarding physical well-being, attention has been directed towards UGIs as a place for outdoor sports activities (i.e., running, walking, biking), but also for relaxation and meditation (i.e., forest bathing and therapy). Considering social inclusion, participants evidenced that the UGIs must be places of aggregation and inclusion, especially for the elderly, children, and people with disabilities.

Finally, the role of UGIs as areas of urban biodiversity emerged as a shared point among the participants at the Scientific Café. This is an important aspect that must be taken into consideration in the management of UGIs, in order to promote some urban wildlife and vegetation diversity. To this end, the management aspects to be considered are promoting ecological connectivity between UGIs and other natural ecosystems, maintaining tree-related microhabitats, and improving diversity in terms of tree, shrub and herbaceous species.

4. Discussion

The results of the present study highlight that the international literature has placed greater importance—in terms of empirical studies conducted worldwide—on the regulating services provided by UGIs (i.e., climate change mitigation, heat islands mitigation), followed by supporting services (i.e., urban biodiversity conservation) and cultural services (i.e., public health, aesthetic value, recreational opportunities). Provisioning services have been the least widely investigated by the scientific community, being limited to the production of timber/biomass and food. This order of relevance among ESs categories was partly confirmed by the discussion with Italian stakeholders during the Scientific Café. The focus of the discussion among the participants was first on cultural and regulating services closely connected with the health and quality of life of citizens, followed by the conservation of the natural environment and biodiversity. Conversely, provisioning services provided by UGIs were not mentioned by any of the participants at the Scientific Café.

The comparison between the importance assigned to ES by the scientific debate and by the social perspective according to the MEA (2005) [16] framework is summarized in Table 3. The MEA framework developed in 2010 represents a turning point in the study of ESs at the international level. As highlighted by our literature review, since 2010, the number of scientific articles on ecosystem services provided by natural resources in general and by UGIs in particular has seen a sharp increase. The MEA Report has stimulated the scientific community to analyze nature's ecosystem services from different perspectives on the one hand, and has raised awareness among policymakers and ordinary citizens on the benefits provided by nature on the other. More recently, IPBES Report (2022) [39] has challenged policy makers with a set of key messages on how to assess nature based on the multiple values of different stakeholders and institutions. Including multiple different perspectives is key to achieving the objectives of the 2050 Vision for Biodiversity and the 2030 Agenda for Sustainable Development [39]. In this political context, our study sought to provide a comparison between the perspectives of the scientific community and those of the technicians and practitioners who manage UGIs.

Table 3. Comparison between the importance assigned to ES by the scientific community and by the general public (++++ = very high importance, +++ = high importance, ++ = medium importance, + = low importance, o = negligible importance).

Ecosystem Services	Scientific Debate	Social Perception
	Provisioning services	
Food production	++	o
Timber and biomass production	+	o
	Regulating services	
Heat islands mitigation	+++	++++
Climate change/Carbon sequestration	++++	++
Water management	++	o
	Cultural services	
Public health, well-being and quality of life	+++	++++
Aesthetic value	+++	+++
Outdoor recreation	+++	+++
	Supporting services	
Urban biodiversity (flora and fauna)	++++	+++

The results of our study reveal that the strongest point of convergence between the quantitative and qualitative data, that is, between the scientific literature and the stakeholders' perspectives, is the role of UGIs in promoting the health, well-being and quality of life of the urban population. Our sample of respondents emphasized that improving the mental and psychological health of citizens is the most important contribution of UGIs, especially in large cities. Participants in the Scientific Café also underlined that UGIs should primarily be managed with a focus on the psychophysical wellbeing of cities' communities. In fact, physical and mental health is closely linked to the quantity and quality of UGIs.

In other words, UGIs that foster individual reflection and social relations are conducive to people's mental well-being, while UGIs that enable physical activities are conducive to people's physical well-being.

In the literature, some authors have studied the relationship between different types of UGIs and people's mental well-being, highlighting that access to UGIs can reduce stress associated with loneliness by providing opportunities to both cultivate personal relationships and engage in community activities [40–42]. In addition, UGIs are perceived as places to cultivate personal well-being, both physically through the practice of outdoor sports (e.g., jogging, walking, biking) and psychologically through relaxation in a natural environment [43]. Regarding this last aspect, a systematic literature review highlighted that exposure to green areas leads to a reduction in stress, a positive mood, and less depressive symptoms [44]. Moreover, numerous studies have emphasized the growing awareness of citizens of the role of UGIs for their effects on physical and mental well-being following the COVID-19 pandemic. A further critical aspect highlighted by the international literature is the methodological difficulty of mapping and evaluating cultural ecosystem services due to the fact that they are intangible benefits [45,46]. In fact, these benefits are firstly subjective, as they are linked to the sphere of individual values, and secondly dependent on the socio-cultural context [47]. Therefore, studies like this one provide empirical data related to a specific context, and can be of help in the creation of a global database useful for highlighting differences and convergences between categories of stakeholders and countries.

In Italy, three empirical studies conducted between 2020 and 2022 found that the restrictions during the COVID-19 pandemic waves in Italy positively influenced citizens' perceptions of UGIs, leading to a greater interest in green areas [2,13,48].

Furthermore, those authors found an increased need to frequent parks and gardens close to their home during the COVID-19 lockdown. Similar findings were reported for other European cities (see Derks et al. and Beckmann-Wübbelt et al. for the cities of Bonn and Karlsruhe in Germany [49,50], Venter et al. for Oslo in Norway [51], and Da Schio et al. for Brussels in Belgium [52]).

Conversely, Lopez et al., in a study conducted in New York City, found that the numbers of citizens who attended UGIs during the COVID-19 pandemic remained unchanged [53]. Although most of these studies were conducted during or shortly after the period of the COVID-19 pandemic restrictions, it is important to note that public awareness of the mental and physical health benefits of UGIs has persisted, as evidenced by the results of our Scientific Café.

Another cultural service emphasized by both the scientific community and participants in the Scientific Café is the aesthetic landscape value related to the presence of UGIs. According to our sample of stakeholders, this service was considered less relevant than the previous ones, but the literature has attributed considerable importance to this theme.

In fact, Dushkova et al. assessed the attractiveness of UGIs (urban parks) in Russia and China, considering among the various factors the aesthetic value of these green areas [54], while He et al. estimated the aesthetic landscape value of urban forest parks in Canada based on the Five Senses Theory [55]. As highlighted by several authors, the size and features of UGIs—such as tree species composition and structure—are key factors that influence citizens' aesthetic perception [56,57]. In line with these findings, our sample of stakeholders emphasized the importance of urban and gardens parks, along with the diversification of structural and tree species diversification. Additionally, they suggested that restoring and redeveloping abandoned infrastructure areas, such as industrial buildings and ruins, through green spaces would be a valuable strategy.

Regarding regulating services, our results show that, according to our sample of Italian stakeholders, the most important ESs provided by UGIs involve the regulation of microclimate and heat. Conversely, climate change mitigation and runoff reduction were underemphasized by the participants at the Scientific Café. This represents the main divergence between the opinions of our sample and the literature on the subject. In fact, several studies have investigated and emphasized the capacity and main role of UGIs in mitigating climate change through the reduction in atmospheric CO_2 [10,58,59].

As highlighted by the international literature, UGIs influence the urban microclimate through temperature regulation and carbon storage [60]. Many studies have also investigated the ability of different types of UGIs to reduce air temperature through the evaporation of water from surfaces (evaporation) or plants (transpiration) [61].

Finally, our results underline that urban biodiversity is a key ecosystem service provided by UGIs, according to Italian stakeholders. Specifically, our sample emphasized the role of UGIs as habitat and food sources for urban wildlife. In the literature, some studies have investigated the role of UGIs as an ecological network that promotes urban habitats and supports biodiversity, or acts as a refuge and habitat for target species of the urban environment [62–64]. To foster and maintain urban biodiversity, it is necessary to pay particular attention to the presence of a network of UGIs, ensuring adequate size, and carefully selecting tree and shrub species.

Our results regarding the importance of ESs assigned by the stakeholders are comparable with the results of other studies conducted with in-depth interviews, focus groups or scientific cafés. In a study conducted in Germany, Riechers et al. [65] interviewed three categories of stakeholders (professionals in planning and decision-making positions at three different institutions) and highlighted the key importance of some cultural ecosystem services, such as the aesthetic beauty of the landscape, psychological well-being, and sense of place through nature. Through focus groups, Kičić et al. [45] investigated the importance of cultural ecosystem services provided by the UGIs of the city of Zagreb. Those authors highlighted that, according to the citizens' opinions, the most important ESs are place attachment, aesthetic and recreation. In another study, Slovák et al. [66] analyzed the studies that used the focus group methodology to collect qualitative information on the importance of cultural ESs. Those authors found that the most investigated cultural ESs are recreation, aesthetics, education, and spiritual value. Palacios-Agundez et al. [67] investigated the importance assigned by stakeholders in the Basque Country (Spain) to ESs using the World Café methodology. Those authors found that the four most important perceived ESs were air quality regulation, water regulation, biodiversity, and environmental education. However, that study used a similar methodology to ours, but considered ESs in general and not those provided by UGIs specifically. In summary, the international literature on the social demand for ESs provided by UGIs has highlighted the high importance of cultural services compared to other categories (i.e., provisioning, regulating, supporting), as highlighted by our study.

5. Conclusions

The present study adopted a mixed qualitative–quantitative case study approach to investigate whether stakeholders' prioritizations of ESs provided by UGIs are in line with those of international research. This approach is potentially useful in providing advice to decision-makers to guide future UGIs planning. On one hand, the results of scientific research offer support in the management of UGIs to improve the provision of key ESs in a given context (e.g., improving air quality, providing outdoor recreation services, improving urban microclimates by providing cooling). On the other hand, stakeholders' perceptions and preferences are important in better directing decision-makers' choices in satisfying social demands. In particular, UGI managers and planners can enhance the most important ESs in the eyes of stakeholders and citizens, thus increasing the social acceptance of choices and raising mutual trust between policymakers and local communities.

Our results show that most of the ESs studied by the scientific community also coincide with those considered most important by stakeholders, such as urban biodiversity conservation, heat islands mitigation, and the improvement of the public health and well-being of citizens. However, some others—such as climate change mitigation through carbon storage and sequestration—have attracted more attention in the international scientific community.

The main hypothesis of this divergence is that the scientific community is interested in investigating the innovative aspects of a topic (e.g., ESs rarely investigated in the past literature) or those with high relevance at the international political level (e.g., international climate change policy), while ordinary people are more interested in everyday aspects of life such as using UGIs for recreational, aesthetic or wellness purposes. A second possible hypothesis regarding the divergence between the scientific community and stakeholders concerning the role of UGIs in climate change mitigation is the complexity of understanding how forests and green infrastructures contribute to climate change mitigation, particularly in terms of carbon storage and sequestration. These processes are indirect and long-term, making them harder for the public to perceive, compared to more immediate benefits like temperature regulation or recreational spaces. While the role of UGIs in mitigating climate change is well-documented in scientific research, it may seem too abstract a concept to stakeholders without a scientific background. Therefore, raising awareness about the long-term environmental benefits of UGIs, through education and targeted communication, could help bridge this gap.

From a methodological point of view, the main strength of this study is in providing an approach that integrates a quantitative analysis of the literature based on BNA with stakeholder involvement through participation, which is capable of collecting qualitative information. Conversely, the main weakness of the present study is having involved local stakeholders from only one country in a single participatory event. Another limitation of the study is presumably related to the composition of the sample of stakeholders who participated in the Scientific Café. In our sample, there were no representatives of the industrial production sector, and this fact may have influenced the negligible importance assigned to provisioning services (food, biomass and timber production). Conversely, the high number of public administrations representatives may have further emphasized the importance of services related to citizens' well-being and quality of life.

Future steps will be to extend the research to other EU countries involved in the ForestValue2 project and to organize a cycle of Scientific Cafés following the same standard procedure in different cities in each country.

Author Contributions: Conceptualization, I.D.M. and A.P.; methodology, I.D.M., A.P. and S.B.; software, A.P.; investigation, S.B.; data curation, A.P. and S.B.; writing—original draft preparation, I.D.M., A.P., S.B. and M.G.C.; supervision, M.G.C. All authors have read and agreed to the published version of the manuscript.

Funding: This research received no external funding.

Data Availability Statement: The data presented in this study are available on request from the corresponding author.

Acknowledgments: The work was supported by the ForestValue2 Project "Innovating forest-based bioeconomy" (Horizon Europe GA no. 101094340), which runs from January 2023 until December 2027.

Conflicts of Interest: The authors declare no conflicts of interest.

References

1. Reyes-Riveros, R.; Altamirano, A.; De La Barrera, F.; Rozas-Vásquez, D.; Vieli, L.; Meli, P. Linking public urban green spaces and human well-being: A systematic review. *Urban For. Urban Green.* **2021**, *61*, 127105. [CrossRef]
2. De Meo, I.; Alfano, A.; Cantiani, M.G.; Paletto, A. The Impact of the COVID-19 Pandemic on Citizens' Attitudes and Behaviors in the Use of Peri-Urban Forests: An Experience from Italy. *Sustainability* **2023**, *15*, 2852. [CrossRef]
3. WHO. *Strengthening Preparedness for COVID-19 in Cities and Urban Settings: Interim Guidance for Local Authorities*; World Health Organization (WHO): Geneva, Switzerland, 2020.
4. Capotorti, G.; De Lazzari, V.; Orti, M.A. Local Scale Prioritisation of Green Infrastructure for Enhancing Biodiversity in Peri-Urban Agroecosystems: A Multi-Step Process Applied in the Metropolitan City of Rome (Italy). *Sustainability* **2019**, *11*, 3322. [CrossRef]
5. Cantiani, M.G.; Betta, A.; De Meo, I.; Maino, F.; Paletto, A.; Tamanini, S. Green cities for a better future?—A case study from an Alpine region: The town of Trento. *Upland* **2019**, *4*, 41–52.
6. Hansen, R.; Mattes, A.; Meier, M.; Kurths, A. Reorienting urban green infrastructure planning towards biodiversity—Perspectives and ongoing debates from Germany. *Urban For. Urban Green.* **2023**, *90*, 128155. [CrossRef]

7. Zulian, G.; Marando, F.; Vogt, P.; Barbero Vignola, G.; Babí Almenar, J.; Zurbarán-Nucci, M.; Prince, K. *BiodiverCities: A Roadmap to Enhance the Biodiversity and Green Infrastructure of European Cities by 2030, EUR 31224 EN*; Publications Office of the European Union: Luxembourg, 2022; JRC129888. [CrossRef]
8. D'Onofrio, R.; Brownlee, T.; Camaioni, C.; Caprari, G.; Malavolta, S.; Trusiani, E. Learning from Experience to Build Urban Green Infrastructure (UGI) in the Central Adriatic City (Italy) under the Life+A_GreeNet Project. *UPLanD* **2023**, *7*, 5–24. [CrossRef]
9. European Commission. *Directorate-General for Environment, Building a Green Infrastructure for Europe*; Publications Office: Luxembourg, 2014. [CrossRef]
10. Gómez-Villarino, M.T.; Gómez-Villarino, M.; Ruiz-Garcia, L. Implementation of Urban Green Infrastructures in Peri-Urban Areas: A Case Study of Climate Change Mitigation in Madrid. *Agronomy* **2021**, *11*, 31. [CrossRef]
11. Esch, T.; Bachofer, F.; Heldens, W.; Hirner, A.; Marconcini, M.; Palacios-Lopez, D.; Roth, A.; Üreyen, S.; Zeidler, J.; Dech, S.; et al. Where We Live—A Summary of the Achievements and Planned Evolution of the Global Urban Footprint. *Remote Sens.* **2018**, *10*, 895. [CrossRef]
12. European Environment Agency. *The European Environment-State and Outlook 2020: Knowledge for Transition to a Sustainable Europe*; European Environment Agency (EEA): Copenhagen, Denmark, 2020; Volume 60, pp. 391–394.
13. De Meo, I.; Becagli, C.; Cantiani, M.G.; Casagli, A.; Paletto, A. Citizens' use of public urban green spaces at the time of the COVID-19 pandemic in Italy. *Urban For. Urban Green.* **2022**, *77*, 127739. [CrossRef]
14. Barton, J.; Pretty, J. What is the best dose of nature and green exercise for improving mental health? A multi-study analysis. *Environ. Sci. Technol.* **2010**, *44*, 3947–3955. [CrossRef]
15. Larabi, F.; Berrichi, M.; Paletto, A. Social Demand for Ecosystem Services Provided by Peri-Urban Forests: The Case Study of the Tlemcen Forest (Algeria). *J. Environ. Account Manag.* **2021**, *9*, 19–29. [CrossRef]
16. MEA. *Ecosystems and Human Well-Being: Biodiversity Synthesis*; World Resources Institute: Washington, DC, USA, 2005.
17. Ouyang, X.; Luo, X. Models for Assessing Urban Ecosystem Services: Status and Outlooks. *Sustainability* **2022**, *14*, 4725. [CrossRef]
18. Zou, Z.; Yang, Y.; Qiu, G.Y. Quantifying the Evapotranspiration Rate and Its Cooling Effects of Urban Hedges Based on Three-Temperature Model and Infrared Remote Sensing. *Remote Sens.* **2019**, *11*, 202. [CrossRef]
19. Uribe, C.; Brenes, R.; Hack, J. Potential of retrofitted urban green infrastructure to reduce runoff—A model implementation with site-specific constraints at neighborhood scale. *Urban For. Urban Green.* **2022**, *69*, 127499. [CrossRef]
20. Del Serrone, G.; Peluso, P.; Moretti, L. Evaluation of Microclimate Benefits Due to Cool Pavements and Green Infrastructures on Urban Heat Islands. *Atmosphere* **2022**, *13*, 1586. [CrossRef]
21. Wu, Q.; Huang, Y.; Irga, P.; Kumar, P.; Li, W.; Wei, W.; Shon, H.K.; Lei, C.; Zhou, J.L. X Synergistic control of urban heat island and urban pollution island effects using green infrastructure. *J. Environ. Manag.* **2022**, *370*, 122985. [CrossRef]
22. Pen, L.L.H.; Jim, C.Y. Economic evaluation of green-roof environmental benefits in the context of climate change: The case of Hong Kong. *Urban For. Urban Green.* **2015**, *14*, 554–561. [CrossRef]
23. Hekrle, M.; Liberalesso, T.; Macháč, J.; Matos Silva, C. The economic value of green roofs: A case study using different cost–benefit analysis approaches. *J. Clean. Prod.* **2023**, *413*, 137531. [CrossRef]
24. Giannico, V.; Spano, G.; Elia, M.; D'Este, M.; Sanesi, G.; Lafortezza, R. Green spaces, quality of life, and citizen perception in European cities. *Environ. Res.* **2021**, *196*, 110922. [CrossRef] [PubMed]
25. Ostoić, S.K.; Konijnendijk van den Bosch, C.C.; Vuletić, D.; Stevanov, M.; Živojinović, I.; Mutabdžija-Bećirović, S.; Lazarević, J.; Stojanova, B.; Blagojević, D.; Stojanovska, M.; et al. Citizens' perception of and satisfaction with urban forests and green space: Results from selected Southeast European cities. *Urban For. Urban Green.* **2017**, *23*, 93–103. [CrossRef]
26. Molari, M.; Dominici, L.; Manso, M.; Matos Silva, C.; Comino, E. A socio-ecological approach to investigate the perception of green walls in cities: A comparative analysis of case studies in Turin and Lisbon. *Nat. Based Solut.* **2024**, *6*, 100175. [CrossRef]
27. Kim, M.; Rupprecht, C.D.D.; Furuya, K. Residents' Perception of Informal Green Space—A Case Study of Ichikawa City, Japan. *Land* **2018**, *7*, 102. [CrossRef]
28. Chen, S.; Wang, Y.; Ni, Z.; Zhang, X.; Xia, B. Benefits of the ecosystem services provided by urban green infrastructures: Differences between perception and measurements. *Urban For. Urban Green.* **2020**, *54*, 126774. [CrossRef]
29. Otte, E.; Rousseau, R. Social network analysis: A powerful strategy, also for the information sciences. *J. Inf. Sci.* **2002**, *28*, 441–453. [CrossRef]
30. Biancolillo, I.; Paletto, A.; Bersier, J.; Keller, M.; Romagnoli, M. A literature review on forest bioeconomy with a bibliometric network analysis. *J. For. Sci.* **2020**, *66*, 265–279. [CrossRef]
31. Van Eck, N.J.; Waltman, L. Visualizing bibliometric networks. In *Measuring Scholarly Impact: Methods and Practice*; Ding, Y., Rousseau, R., Wolfram, D., Eds.; Springer: Cham, Switzerland, 2014; pp. 285–332.
32. Nesseth, N.M.; Henson, A.M.; Barriault, C.L. A Framework for Understanding the Nature of Questions Asked by Audience Participants at Science Cafés. *Front. Educ.* **2021**, *6*, 674878. [CrossRef]
33. Dijkstra, A.M. Analysing Dutch Science Cafés to better understand the science-society relationship. *J. Sci. Commun.* **2017**, *16*, 1–17. [CrossRef]

34. Baldessari, S.; Paletto, A.; De Meo, I. Rethinking Public Participation in Forest Policies: A Literature Review of Participatory Techniques. *Forests* **2024**, *15*, 1514. [CrossRef]
35. De Meo, I.; Apostol, E.N.; Kallio, M.; Marongiu, S.; Badea, O.N.; Paletto, A. Scientific Café as a tool to involve civil society in the forestry sector: A methodological approach. *Agric. For. Podgor.* **2023**, *69*, 285–301. [CrossRef]
36. Grilli, G.; Garegnani, G.; Poljanec, A.; Ficko, A.; Vettorato, D.; De Meo, I.; Paletto, A. Stakeholder analysis in the biomass energy development based on the experts' opinions: The example of Triglav National Park in Slovenia. *Folia For. Pol.* **2015**, *57*, 173–186. [CrossRef]
37. Mitchell, R.; Agle, B.; Wood, D. Towards a theory of stakeholder identification: Defining the principle of who and what really counts. *Acad. Manag. Rev.* **1997**, *22*, 853–886. [CrossRef]
38. Long, A.J.; Nair, P.K.R. Trees outside forests: Agro-, community, and urban forestry. *New For.* **1999**, *17*, 145–174. [CrossRef]
39. IPBES. Summary for Policymakers of the Methodological Assessment Report on the Diverse Values and Valuation of Nature of the Intergovernmental Science-Policy Platform on Biodiversity and Ecosystem Services. In *Intergovernmental Science-Policy Platform on Biodiversity and Ecosystem Services (IPBES) Secretariat*; Pascual, U., Balvanera, P., Christie, M., Baptiste, B., González-Jiménez, D., Anderson, C.B., Athayde, S., Barton, D.N., Chaplin-Kramer, R., Jacobs, S., et al., Eds.; IPBES: Bonn, Germany, 2022; p. 45. [CrossRef]
40. Marques, P.; Santos Silva, A.; Quaresma, Y.; Resende Manna, L.; de Magalhães Neto, N.; Mazzoni, R. Home gardens can be more important than other urban green infrastructure for mental well-being during COVID-19 pandemics. *Urban For. Urban Green.* **2021**, *64*, 127268. [CrossRef]
41. Jabbar, M.; Mohd Yusoff, M.; Shafie, A. Assessing the role of urban green spaces for human well-being: A systematic review. *GeoJournal* **2021**, *87*, 4405–4423. [CrossRef]
42. Pasanen, T.P.; White, M.P.; Elliott, L.R.; van den Bosch, M.; Bratman, G.N.; Ojala, A.; Korpela, K.; Fleming, L.E. Urban green space and mental health among people living alone: The mediating roles of relational and collective restoration in an 18-country sample. *Environ. Res.* **2023**, *232*, 116324. [CrossRef]
43. Vujcic, M.; Tomicevic-Dubljevic, J.; Zivojinovic, I.; Toskovic, O. Connection between urban green areas and visitors' physical and mental well-being. *Urban For. Urban Green.* **2019**, *40*, 299–307. [CrossRef]
44. Zhang, Y.; Mavoa, S.; Zhao, J.; Raphael, D.; Smith, M. The Association between Green Space and Adolescents' Mental Well-Being: A Systematic Review. *Int. J. Environ. Res. Public Health* **2020**, *17*, 6640. [CrossRef]
45. Kičić, M.; Haase, D.; Marin, A.M.; Vuletić, D.; Krajter Ostoić, S. Perceptions of cultural ecosystem services of tree-based green infrastructure: A focus group participatory mapping in Zagreb, Croatia. *Urban For. Urban Green.* **2022**, *78*, 127767. [CrossRef]
46. Baumeister, C.F.; Gerstenberg, T.; Plieninger, T.; Schraml, U. Exploring cultural ecosystem service hotspots: Linking multiple urban forest features with public participation mapping data. *Urban For. Urban Green.* **2020**, *48*, 126561. [CrossRef]
47. Kim, J.; Son, Y. Assessing and mapping cultural ecosystem services of an urban forest based on narratives from blog posts. *Ecol. Indic.* **2021**, *129*, 107983. [CrossRef]
48. Larcher, F.; Pomatto, E.; Battisti, L.; Gullino, P.; Devecchi, M. Perceptions of Urban Green Areas during the Social Distancing Period for COVID-19 Containment in Italy. *Horticulturae* **2021**, *7*, 55. [CrossRef]
49. Derks, J.; Giessen, L.; Winkel, G. COVID-19-induced visitor boom reveals the importance of forests as critical infrastructure. *For. Policy Econ.* **2020**, *118*, 102253. [CrossRef]
50. Beckmann-Wübbelt, A.; Fricke, A.; Sebesvari, Z.; Yakouchenkova, I.A.; Fröhlich, K.; Saha, S. High public appreciation for the cultural ecosystem services of urban and peri-urban forests during the COVID-19 pandemic. *Sustain. Cities Soc.* **2021**, *74*, 103240. [CrossRef]
51. Venter, Z.S.; Barton, D.N.; Gundersen, V.; Figari, H.; Nowell, M. Urban nature in a time of crisis: Recreational use of green space increases during the COVID-19 outbreak in Oslo, Norway. *Environ. Res. Lett.* **2020**, *15*, 104075. [CrossRef]
52. Da Schio, N.; Phillips, A.; Fransen, K.; Wolff, M.; Haase, D.; Krajter Ostoić, S.; Živojinović, I.; Vuletić, D.; Derks, J.; Davies, C.; et al. (The impact of the COVID-19 pandemic on the use of and attitudes to-wards urban forests and green spaces: Exploring the instigators of change in Belgium. *Urban For. Urban Green.* **2021**, *65*, 127305. [CrossRef]
53. Lopez, B.; Kennedy, C.; Field, C.; McPhearson, T. Who benefits from urban green spaces during times of crisis? Perception and use of urban green spaces in New York City during the COVID-19 pandemic. *Urban For. Urban Green.* **2021**, *65*, 127354. [CrossRef]
54. Dushkova, D.; Ignatieva, M.; Konstantinova, A.; Yang, F. Cultural Ecosystem Services of Urban Green Spaces. How and What People Value in Urban Nature? In *Advanced Technologies for Sustainable Development of Urban Green Infrastructure*; Vasenev, V., Dovletyarova, E., Valentini, R., Cheng, Z., Calfapietra, C., Inostroza, C., Leuchner, M., Eds.; Springer: Cham, Switzerland, 2021; pp. 292–318. [CrossRef]
55. He, M.; Wang, Y.; Wang, W.J.; Xie, Z. Therapeutic plant landscape design of urban forest parks based on the Five Senses Theory: A case study of Stanley Park in Canada. *Int. J. Geoheritage Parks* **2022**, *10*, 97–112. [CrossRef]
56. Rey Gonzalo, G.; Barrigón Morillas, J.M.; Montes González, D. Perceptions and use of urban green spaces on the basis of size. *Urban For. Urban Green.* **2019**, *46*, 126470. [CrossRef]

57. Paletto, A.; Guerrini, S.; De Meo, I. Exploring visitors' perceptions of silvicultural treatments to increase the destination attractiveness of peri-urban forests: A case study in Tuscany Region (Italy). *Urban For. Urban Green.* **2017**, *27*, 314–323. [CrossRef]
58. Parsa, V.A.; Salehi, E.; Yavari, A.R.; van Bodegom, P.M. Evaluating the potential contribution of urban ecosystem service to climate change mitigation. In *Urban Ecosystems*; Jim, C.Y., McKinney, M., Eds.; Springer Science + Business Media: Berlin, Germany, 2019; pp. 989–1006. [CrossRef]
59. keobilor, J. Urban Green Infrastructure and its effects on Climate Change—A Review. *SSRN Electron. J.* **2022**. [CrossRef]
60. Artmann, M.; Inostroza, L.; Fan, P. Urban sprawl, compact urban development and green cities. How much do we know, how much do we agree? *Ecol. Indic.* **2018**, *96*, 3–9. [CrossRef]
61. Reinwald, F.; Brandenburg, C.; Gabor, A.; Hinterkoerner, P.; Kainz, A.; Kraus, F.; Ring, Z.; Scharf, B.; Toetzer, T.; Damyanovic, D. Multi-level toolset for steering urban green infrastructure to support the development of climate-proofed cities. *Sustainability* **2021**, *13*, 12111. [CrossRef]
62. Minor, E.S.; Anderson, E.C.; Belaire, J.A.; Garfinkel, M.; Smith, A.D. Urban green infrastructures and ecological networks for urban biodiversity conservation. In *Urban Biodiversity*; Ossola, A., Niemelä, J., Eds.; From Research to Practice: London, UK; Routledge: London, UK, 2017; pp. 186–199.
63. Hansen, R.; Olafsson, A.S.; van der Jagt, A.P.N.; Rall, E.; Pauleit, S. Planning multifunctional green infrastructure for compact cities: What is the state of practice? *Ecol. Indic.* **2019**, *96*, 99–110. [CrossRef]
64. Liu, Z.; Zhou, Y.; Yang, H.; Liu, Z. Urban green infrastructure affects bird biodiversity in the coastal megalopolis region of Shenzhen city. *Appl. Geogr.* **2023**, *151*, 102860. [CrossRef]
65. Riechers, M.; Barkmann, J.; Tscharntke, T. Perceptions of cultural ecosystem services from urban green. *Ecosyst. Serv.* **2016**, *17*, 33–39. [CrossRef]
66. Slovák, L.; Daněk, J.; Daněk, T. The use of focus groups in cultural ecosystem services research: A systematic review. *Humanit. Soc. Sci. Commun.* **2023**, *10*, 45. [CrossRef]
67. Palacios-Agundez, I.; Casado-Arzuaga, I.; Madariaga, I.; Onaindia, M. The Relevance of Local Participatory Scenario Planning for Ecosystem Management Policies in the Basque Country, Northern Spain. *Ecol. Soc.* **2013**, *18*, 7. [CrossRef]

land

Article

Experts' Perspectives on Private Forest Owners' Priorities and Motivations for Voluntary Ecosystem Protection in Lithuania

Diana Lukmine * and Stasys Mizaras

Lithuanian Research Centre for Agriculture and Forestry, LT-58344 Akademija, Lithuania; stasys.mizaras@lammc.lt
* Correspondence: diana.lukmine@lammc.lt

Abstract: Lithuania has initiated the development of voluntary ecosystem protection measures within private forests, establishing protection agreements between the state and private forest owners. This article examines the priorities and motivations of private forest owners in the voluntary protection of ecosystems, based on the analysis of expert opinions. The Delphi sociological method was employed to assess expert opinions on the priorities and motivations of private forest owners regarding the voluntary protection of ecosystems. Twenty-nine experts responded to the survey, providing insights into the attitudes of Lithuanian private forest owners towards voluntary forest protection models and contract types, potential environmental protection instruments, the necessity of compensation for losses incurred due to forest management restrictions in protected areas, the proportion of protected forests, factors influencing the intention to engage in forest protection, motivations for voluntary forest protection, the "crowd-out" effect, sources of compensation for losses, the effectiveness of ecosystem protection mechanisms in Lithuanian forests, and the factors that diminish their effectiveness. Summarizing the experts' findings, it can be concluded that the forest protection priorities of Lithuanian private forest owners, concerning the expansion of protected areas in private forests, protection models, and incentives for protection, are likely to align with the priorities and motivations identified in other European countries. A heterogeneity of priorities and motives was identified. Almost three-quarters of experts thought the current amount of protected forest in Lithuania is sufficient or is already more than necessary, and only about one in ten thought that is necessary for owners to protect more forest. Lithuanian private forest owners are mostly motivated by full financial compensation for losses. According to experts, the majority of private forest owners do not support forest protection models that lack financial compensation. It would be appropriate to implement both permanent and fixed-term protection agreements (contracts) with compensation, alongside the option of selling forests to the state. The level of compensation is identified as the most-significant factor influencing private forest owners' willingness to engage in ecosystem protection. Experts highlight that the primary reasons for the ineffectiveness of private forest protection measures in Lithuania include inadequate and unjustified compensation, compensation amounts that are too low relative to forest owners' income, insufficient information, and complex bureaucratic procedures.

Keywords: priorities; motives; private forest owners; voluntary protection; ecosystems

1. Introduction

Protected areas (PAs) are often established to protect biodiversity. In Lithuania, private forests encompass 957.9 thousand hectares, with 24% designated as protected areas.

These protected areas are established mandatorily, and private forest owners often receive incomplete or no compensation for economic restrictions imposed by such designations. Lithuania has implemented ecosystem protection measures, including the safeguarding of key habitats and the designation of Natura 2000 sites. The introduction of voluntary and contractual forest protection is anticipated to expand protected areas and facilitate compensation for forest owners facing economic activity restrictions, particularly in logging. The concept of voluntary forest protection has attracted significant interest since the early 2000s in Finland, Sweden, Norway [1], and other European countries [2]. In Lithuania, legal prerequisites have been created for the conclusion of protection agreements between the state and private forest owners (Republic of Lithuania Law on Protected Areas, 2024). This creates the conditions for more voluntary protection of ecosystems in the country's private forests. Consequently, the first step in designing effective voluntary protection mechanisms is to analyze the priorities and motivations of private forest owners to engage in ecosystem protection voluntarily. These priorities and motives are revealed through surveys or implementing forest protection programs. A survey of private forest owners has been conducted in European countries regarding the priority and motive of voluntary forest protection. Table 1 presents the list of articles about these surveys.

Table 1. List of articles about private forest owners' priorities and motives for voluntary forest protection in Europe.

Article	Country
Kosenius [3]	Finland
Gundersen et al. [4]	Norway
Baranovskis et al. [5]	Latvia
Koskela and Karppinen [6]	Finland
Juutinen et al. [7]	Finland
Danley et al. [8]	Sweden
Abildtrup et al. [9]	France
Palome [10]	France
Mitani and Lindjeni [11]	Norway
Goritz et al. [12]	Finland, Denmark, France, Germany
Gotmark [13]	Sweden

Kosenius [3] explores the attitudes and preferences of Finnish non-industrial private forest owners for voluntary temporary forest conservation (the terms "conservation" and "protection" are used according to literature sources). According to the survey data, forest owners are willing to simultaneously conserve biodiversity and forest carbon. They prefer non-profit organizations as an implementer of programs, suggesting an alternative to the current implementation of forest conservation by authorities. The interest of forest owners in conservation programs increases with the offer of shorter contracts and higher payments. Forest owners differ in terms of the perceived importance of ecological, economic, and social aspects of the sustainability of forestry. Heterogeneity in attitudes and preferences stems from the size of forest land, gender, amount of free time, having a home located on the forest site, and place of residence. Gundersen et al. [4] indicates the preferences of Norway's private forest owners on the different models of forest protection: (1) Protection process controlled by the authorities where the ownership rights completely go to the state for all future with full financial compensation—3% of respondents. (2) Voluntary protection process when parts of the right of use, for example, the right to cut down forests, are granted to the state for all future with full compensation—7%. (3) Agreement where parts of the right of use, such as the right to harvest, are granted to the state for a limited period, a maximum 50 years, with compensation for lost income during the future period—12%.

(4) An agreement in which parts of the right of use, such as the right to harvest forest, are granted to the state for a limited period, a maximum of 15 years, with compensation considerably lower than the lost income during the future period—3%. (5) Agreement where the state financially compensates the forest owner for environmentally friendly management by a silviculture plan—69%. (6) Voluntary protection of the forest without compensation—6%.

Baranovskis et al. [5] aimed to explore factors driving private forest owners' attitudes towards biodiversity conservation on their land. The results suggested that financial dependence on income from the forest harvest was the main factor decisive of the negative attitude of landowners towards biodiversity conservation measures for private land in Latvia. This factor was significantly correlated with forest owners' education, occupation, and size of forest land. However, general conservation values also influenced landowner willingness to accept conservation measures on private forest land. A survey of forest owners regarding forest biodiversity protection measures in Finland by Koskela and Karppinen [6] revealed that most respondents (27%) expressed interest in fixed-term protection contracts. Additionally, 18% favored permanent protection contracts, 17% were inclined to sell their forests to the state for protection, while only 8% supported fixed-term contracts without compensation, and 7% preferred permanent protection without compensation. Notably, 35% of respondents indicated no interest in protection contracts. Juutinen et al. [7], studying forest owners' preferences for the contract-based protection of natural values, also found a heterogeneity of views among forest owners. It was concluded that most forest owners would like to participate in projects promoting biodiversity and other non-market ecosystem services. A total of 74% of respondents chose at least one of the offered protection options. However, they asked for a sufficiently high compensation for contract-based forest management. A segment of forest owners would agree to enter into contracts for lower compensations. Therefore, promoting new forest management strategies and participation in contractual compensation schemes requires additional information and education. Forest owners prefer shorter-term protection contracts. Forest owners perceive contracts longer than 10 years as an unwanted limitation of their freedom to manage their forests. Danley et al. [8] concluded that relatively few Swedish family forest owners think more biodiversity protection is needed. Almost three-quarters of respondents thought the current amount of protected forest is sufficient (52.8%) or is already more than necessary (20.6%), and only about one in ten thought that it is necessary for owners to protect more forest. One-third of respondents are not necessarily against the increasing government regulation in Swedish environmental policy, even if it may restrict some of their decision-making possibilities. Approximately 1 out of 10 (9%) respondents expressed an interest in voluntarily entering into a contract for permanent or time-limited forest conservation with the state, making additional voluntary forest set-asides, or increasing the quality of the retention trees left during felling. There are three-times as many owners who are generally supportive of environmental regulations compared to owners willing to take additional voluntary measures. This suggests that the potential for relying exclusively on additional volunteerism within the current Swedish Forest Model may yield a marginal benefit. Abildtrup et al. [9] hypothesized that the purely monetary nature of the incentives could cause a "crowding-out" effect, i.e., forest owners may reduce their voluntary contribution to biodiversity protection that is driven by pro-social motivations (altruism, self-image, etc.). Institutional factors are significant for forest owners' commitments to biodiversity protection in their forests. A key question concerning the design of voluntary protection schemes involves carefully choosing the scheme authority. Organizations that forest owners trust and have local knowledge of their forests facilitate forest owners' commitment and will consequently increase the cost-effectiveness of the scheme. Social motivation (reputation) is important for commitment as about half of the forest owners are more likely to commit when it is public, particularly if they know they are the first to commit in their neighborhood. Intrinsic motivation is also an important determinant. The majority of the owners have a moral obligation to protect biodiversity. Besides the importance of the choice of institutions, the main lesson is that voluntary biodiversity protection schemes need to account for the different types of forest owners, namely owners who are intrinsically motivated and would participate without demand for compensation, owners who can be influenced through nudges, e.g., making commitments public, and owners for whom non-participation is a matter of principle. Policymakers must recognize the diversity of forest owners and adapt policies to this reality, e.g., by applying more than one incentive type. Palome [10] investigated the motives of private forest owners for participating in seven of the biodiversity-related protection programs. Economic motives have a clear positive effect on the probability of adopting programs. Social motivation significantly impacts the adoption of three programs, while intrinsic motivation has a clear positive effect only on one program. Mitani and Lindjeni [11] suggested that landowners' positive expectations about additional income opportunities, over and beyond the compensation payment, increases the likelihood that owners will participate in conservation programs. In addition, the owners' positive environmental attitude increases the probability of participation. Owners who perceive the regulations as too strict and have more mature forests in their stands are less likely to participate. Goritz et al. [12] concluded that participation in implementing a voluntary policy instrument for land use management implies motivational requirements of the targeted landowner. Increasing knowledge of the potential economic, managerial, and attitudinal factors helps design incentives in accordance and facilitates effective performance. Participation rates for different schemes aimed at enhancing the provision of ecosystem services were contrasted with a range of landowners' socioeconomic statuses, forest management variables, and the instrument design characteristics. Results show larger participation trends in mechanisms that promote a forest ecosystem service while simultaneously augmenting benefits enjoyed by the landowner. Being involved in a forestry association increases the likelihood of engaging in the policy mechanism, especially for small- and medium-sized landowners. Gotmark [13] concluded that the protection of the lands of non-industrial private forest owners sometimes leads to conflicts. According to the survey of southern Sweden, conflict was reported by 22.5% of respondents, while 14% reported good relations with authorities. The respondents reported conflicts over logging rules and compensation.

Moreover, the priorities and motives of private forest owners are well revealed when they participate in voluntary forest protection programs. The ten-year agreements are most popular in the Forest Biodiversity Programme for Southern Finland (METSO) implemented in 2008. In 2023, almost 10,000 hectares of forest habitats were protected under this program. The ten-year environmental aid agreements concluded by Finnish forest owners covered more than 5100 (51%) hectares of forest and about 4300 (43%) hectares were protected permanently (by private nature reserves and selling the land to the state for conservation purposes). Forest owners value the voluntary approach, the independence in decision making, and the chance to retain their property rights. Forest owners receive full financial compensation equivalent to the value of timber at the protected site. If the forest owner chooses to sell the property to the state for permanent protection, the value of the land will also be compensated. With permanent protection, the private forest owner's compensation is tax-free. The nature management projects come at no cost to the forest owner. Protected and managed sites can also be used for nature-based tourism and recreation. Forest owners are assisted in drawing up environmental forestry subsidy applications [5,14–22]. In Sweden, voluntary nature conservation agreements have been in place since 1993, based on civil law agreements between landowners and the Swedish Forest Agency. These agreements typically last 50 years and outline the purpose, appropriate maintenance measures, and regulations regarding usage restrictions. The Swedish Environmental Protection and Forest Agency jointly proposed a complementary approach to landowners to initiate Voluntary Agreements in 2008. This method was first tested in the pilot project "KOMET" in 2010, which increased the available forest protection funding under Voluntary Agreements. After the trial period, this approach was implemented nationally. According to the Swedish experience, voluntarism within the current Swedish forest policy system is effective when it aligns with market interests in timber production [23–25]. Norway's experience indicates that since 2003 forest protection processes have become predominantly voluntary. Before 2004, PAs were established by the government (there were many conflicts, court cases, and expensive and slow procedures). After 2004, under voluntary protection forest owners submitted their forests for protection (there were reduced conflicts, no more court cases, more-efficient methods, and more funds for protection). As around 75% of Norway's forests are privately owned, the participation of private forest owners is essential for achieving biodiversity protection targets. However, the primary obstacle in this transition is the limited government budget for nature protection efforts [26].

The primary lesson learned from international experience is that economic compensation is essential. Private forest owners are more likely to engage in voluntary protection if they receive adequate financial compensation for the economic restrictions imposed by protection measures. The experiences of Finland [3,6], Sweden [8], and Norway [4,11] demonstrate that comprehensive compensation schemes, including tax benefits, enhance participation rates.

The second lesson learned is that shifting from mandatory to voluntary protection mechanisms reduces conflicts between forest owners and authorities, as observed in Sweden [13].

The third lesson learned is that private forest owners have heterogeneous motivations for protection, including economic, ecological, and social considerations. Policy instruments should be tailored to accommodate different levels of interest and willingness to engage in protection measures [7].

The fourth lesson learned is that the effectiveness of voluntary protection schemes depends on the trust forest owners place in implementing institutions. Local and non-governmental organizations may play a key role in improving credibility and engagement, as suggested by studies from Finland [12] and France [9].

The fifth lesson learned is that many forest owners favor fixed-term agreements (for example, 10-year agreements) over permanent protection commitments [14,15].

The sixth lesson learned is that, beyond financial incentives, factors such as social recognition, reputation, and a sense of moral obligation influence forest owners' willingness to engage in voluntary protection [10].

The attitudes of Lithuanian private forest owners towards voluntary forest protection have not been previously studied. Therefore, this study aimed to identify experts' opinions on the priorities and motivations of Lithuanian private forest owners regarding the voluntary protection of ecosystems.

2. Materials and Methods

The Delphi sociological method was employed to assess experts' perspectives on the priorities and motivations of private forest owners regarding the voluntary protection of ecosystems. This method involves a questionnaire, expert groups, a survey, and an evaluation of results [27,28], with several stages, including expert selection, questionnaire design, discussions, and survey implementation, among others. This study examines three key issues on which experts are expected to reach a consensus: (1) priorities for ecosystem protection models, (2) the types of protection agreements (contracts), and (3) the values that should be compensated when a contract includes provisions for financial compensation.

Compilation of the Questionnaire. The experts were surveyed using a questionnaire designed to gather their opinions on the priorities and motivations of Lithuanian private forest owners in voluntarily protecting ecosystems. The questionnaire was developed after reviewing literature sources [1–26]. These sources included scientific research and analyses on the priorities and motivations for the voluntary preservation of ecosystems in private forests, as well as methods for compensating losses and the effectiveness of various compensation models. The questionnaire consists of 16 closed-ended questions, grouped into three thematic areas: priorities, motivations, and the effectiveness of implemented measures. The first group (Priorities) included questions 1–4, the second group (Motives) included questions 5–14, and the third group (Effectiveness) included questions 15–16. The questions from the questionnaire are presented in the results section of this paper.

Selection of experts. A fundamental criterion of expert group formation is that experts must know the problem under investigation. According to Dalkey [27,28], the optimal number of respondents for an experts' survey is between 25 and 30. For this study, 30 experts were selected and divided into two equal groups, each comprising individuals with expertise in forestry and environmental protection, or with substantial experience as private forest owners.

The first group consisted of 15 expert specialists representing various sectors within forest-based activities, including forestry science, business, management, and public organizations. For the selection of experts, a list of key institutions related to forest ecosystem protection was compiled, with each institution required to include at least one representative possessing the a high level of competence in the subject under study. These experts were selected based on their advanced competence and detailed knowledge pertinent to this research. The selection criteria for participation were as follows: (1) a minimum of a university degree; (2) at least five years of professional experience in forestry, ecology, or environmental protection; and (3) preference was given to individuals holding managerial positions. Other characteristics, such as gender or age, were not deemed critical for expert selection. Most experts held master's or doctoral degrees, while others had higher university qualifications.

The second group comprised expert practitioners, with 15 active and experienced private forest owners. The selection criterion for this group was straightforward: participants were required to be actively engaged in forest management and to have considerable experience as private forest owners. Additionally, they needed to be involved in public organizational activities related to forestry or environmental protection. Both groups of experts were presented with the same set of questions. The response rate to the questionnaire was 96.7%, with one incomplete submission excluded from the data analysis. To ensure respondent confidentiality and encourage candid responses, as required by research ethics [29,30], the questionnaires were coded, ensuring that all responses remained anonymous.

Assessment of the internal consistency of a questionnaire scale. Cronbach's alpha coefficient is used to assess the internal consistency of a questionnaire scale. It is based on the correlation between individual items comprising the questionnaire and evaluates whether all scale items sufficiently reflect the measured construct, while also allowing for the refinement of the required number of items in the scale. Cronbach's alpha coefficient is interpreted as the correlation between the given scale and all other possible scales, comprising the same number of items, that could be used to assess the characteristic of interest. Cronbach's alpha coefficient ranges from 0 to 1. For research purposes, a value of 0.60 is considered acceptable; however, 0.70 or higher is generally desired for the scale or set of items to be deemed consistent [27–29].

3. Results

A total of 30 experts participated in the survey; however, one questionnaire was incomplete and subsequently excluded from the final analysis. After conducting the experts' survey, analyzing the data, and discussing the findings, a consensus was reached on the most-critical issues: (1) the priorities for ecosystem protection models; (2) the types of protection agreements (contracts); and (3) the values to be compensated. Following the discussion of the results, it was concluded that there was no need for experts to revise their responses in subsequent rounds.

The results section of this paper presents data and analysis based on the 29 completed questionnaires. IBM SPSS Statistic version 25 was used for data analysis. The evaluation of the questionnaire's consistency, as measured by Cronbach's alpha coefficient, demonstrates that all three sections are adequately constructed. The Cronbach's alpha value exceeds 0.6 and approaches 0.7, indicating an acceptable level of internal consistency (Table 2).

Table 2. Results of the questionnaire consistency evaluation.

Thematic Parts of Questionnaire	Cronbach's Alpha
Priorities	0.691
Motives	0.695
Effectiveness	0.693

The responses to Question 1, "*Which of the following biodiversity and ecosystem protection models would be most appropriate for implementation in Lithuanian private forests?*", are summarized in Figure 1. Two models were identified as being potentially applicable in Lithuania. The first model involves the state designating PAs for ecosystem protection in private forests, accompanied by compensation. The second model allows private forest owners to voluntarily propose PAs, also with compensation. Both models were supported by 82.8% of experts. Both expert groups agreed that models where PAs are either designated by the state or voluntarily proposed by private forest owners without compensation are not considered viable in Lithuania.

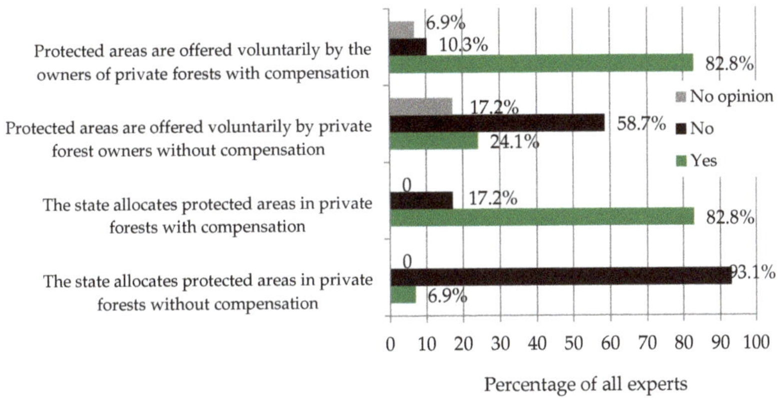

Figure 1. Experts' perspectives on the priorities for ecosystem protection models in Lithuanian private forests.

The responses to Question 2, "*In your opinion, which of the following biodiversity and ecosystem protection models are most likely to be accepted by private forest owners in Lithuania?*", are presented in Figure 2. This question is intended to evaluate the experts' opinions on the attitudes of private forest owners. According to the experts' perspectives, the majority of private forest owners are inclined to support or partially support forest protection measures, provided that appropriate compensation is offered. Private forest owners are likely to disagree or partially disagree with the model in which the state designates protected areas (PAs) in private forests without offering compensation, as indicated by 96.6% of experts. Similarly, 72.4% of experts suppose that private forest owners would not support or would partially oppose a model where PAs are voluntarily offered by private forest owners without compensation. A quantitative survey of private forest owners could provide a completely precise answer to this question.

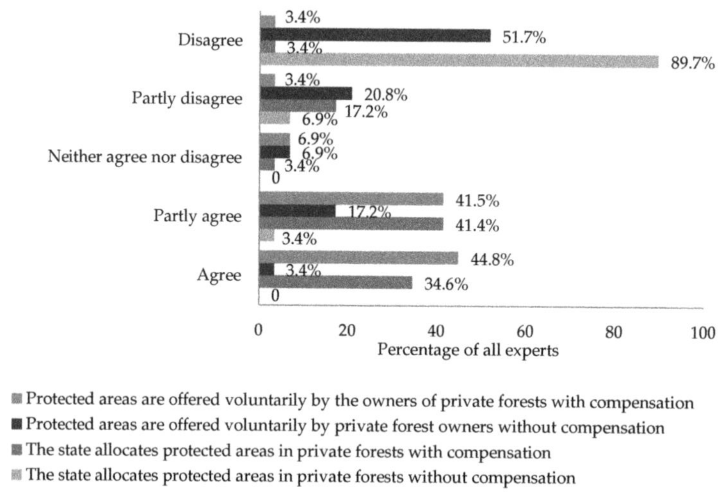

Figure 2. Experts' perspectives on acceptance by private forest owners in Lithuania biodiversity and ecosystem protection models.

The responses to Question 3, "*What types of protection agreements (contracts) should be established with private forest owners in Lithuania?*", are presented in Figure 3. This question presents experts' opinions on the appropriate protection agreements (contracts) to be implemented in Lithuania, while the fourth question aims to evaluate the popularity of different agreements among private forest owners based on experts' assessments. More than 80% of experts indicated that protection agreements (contracts) without compensation should not be implemented in Lithuania. Permanent protection with compensation, while retaining forest ownership, was identified as the most-appropriate type of protection agreement (72.5%). Additionally, selling forests to the state was deemed appropriate by 55.2% of experts.

Additionally, more than 48% of experts suppose that fixed-term contracts of 10 or 20 years, with compensation and without restrictions following the contract's conclusion, could be feasible in Lithuania. Notably, approximately 21% of experts had no opinion on the protection models involving the sale of forests to the state, suggesting that the information regarding protection models and their types is insufficient. Enhancing information dissemination could influence the voluntary protection of private forests.

The responses to Question 4, "*What proportion of private forest owners would be willing to enter into different types of Protection agreements (contracts)?*", are presented in Figure 4. The responses to this question indicate potential trends in the types of contracts that private forest owners would most frequently prefer, based on experts' assessments. Experts indicated that protection agreements (contracts) lacking compensation are unlikely to be favored by private forest owners in Lithuania. A total of 70% of experts suppose that private forest owners are unlikely to be interested in signing fixed 20-year or permanent protection agreements (contracts) without compensation. Furthermore, 55.3% of experts suppose that private forest owners are unlikely to engage in fixed 10-year agreements (contracts) without compensation. A quantitative survey of private forest owners could provide a completely precise answer to this question.

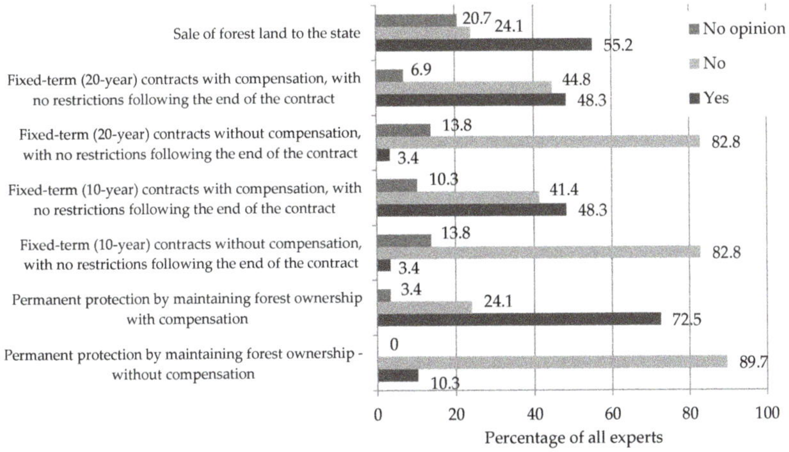

Figure 3. Experts' perspectives on the types of protection agreements (contracts) that could be established with private forest owners in Lithuania.

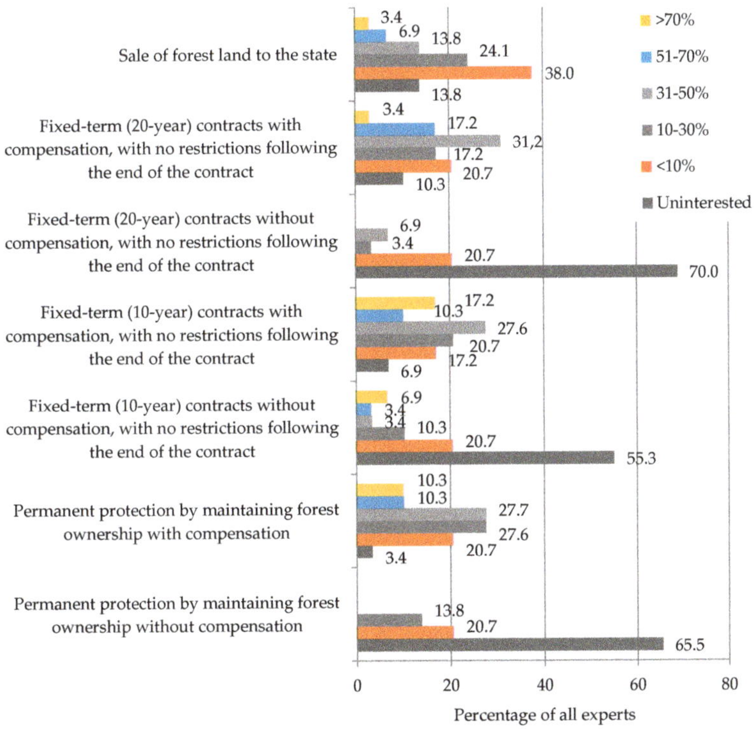

Figure 4. Experts' perspectives on the popularity of protection agreements (contracts) among private forest owners in Lithuania.

The responses to Question 5, *"What are the most important environmental policy instruments for forests?"*, are presented in Table 3. In this instance, the assessment determined which instrument would have the greatest rate of acceptance. Among the three instruments evaluated—land acquisition, protection agreements, and information—experts rated information as the most important, with a mean score of 4.31 (st. dev.—1.317). Protection agreements were also highly rated, with a mean score of 4.17 (st. dev.—1.104). The lowest score was assigned to the instrument "land acquisition", which received a mean score of 3.34 (st. dev.—1.105). Experts consider all three instruments to be very important or extremely important. Consequently, they recommended incorporating all three environmental policy instruments in the development of a new system. Notably, only a minority of experts rated these instruments as unimportant or slightly important.

Table 3. Experts' perspectives on environmental policy instrument acceptance.

Instrument	Score of Likert Scale * Number of Experts/%					Mean Score	St. Dev.
	1	2	3	4	5		
Land acquisition	3/10.3	5/17.2	7/24.3	7/24.1	7/24.1	3.34	1.317
Protection agreements	1/3.4	2/6.9	3/10.3	8/27.7	15/51.7	4.17	1.104
Information	1/3.4	2/6.9	2/6.9	6/20.7	18/62.1	4.31	1.105

* Value of score: 1—not important; 2—slightly important; 3—moderately important; 4—very important; 5—extremely important.

The responses to Question 6, *"What values should be compensated for private forest owners?"*, are presented in Figure 5. All experts agreed that private forest owners should receive compensation for losses incurred due to restrictions on economic activities. Nearly 70% of experts indicated that forest owners should be compensated for lost income and additional expenses. However, over 60% of experts stated that public values should not be subject to compensation.

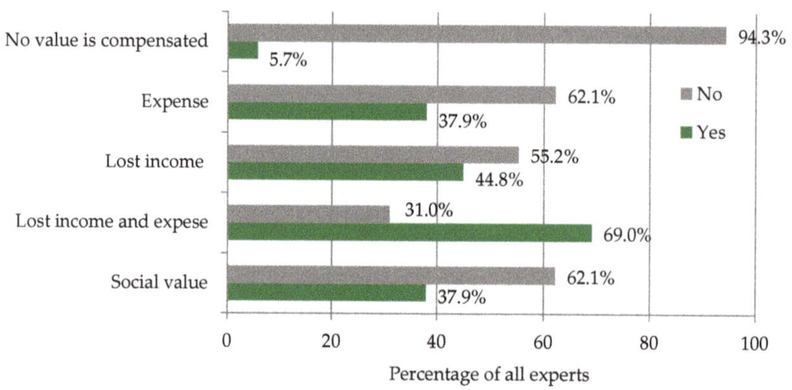

Figure 5. Experts' perspectives on the values that should be compensated to private forest owners.

The responses to Question 7 show experts' opinions on the proportion of private forests that should be designated as protected areas. The survey revealed that 86.2% of experts do not perceive a shortage of protected forests in Lithuania. Furthermore, 55.2% believe that the current proportion—where protected forests constitute 24% of the total forested area—is suboptimal, indicating an excess of protected forests (Figure 6). This consensus among experts suggests a need for reform in Lithuania's approach to forest protection. Notably, four experts abstained from expressing an opinion on this matter.

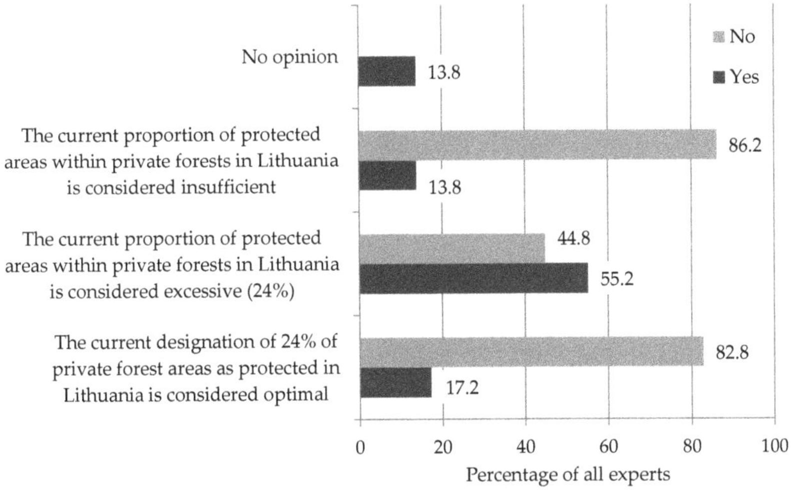

Figure 6. Experts' perspectives on the proportion of protected areas in private forests.

The responses to Question 8, "*Which factors most determine the intention of private forest owners to protect biodiversity and ecosystems?*", are presented in Table 4. The most-determining factors were identified as the amount of compensation (mean score: 4.69), clarity in defining compensation (mean score: 4.52), restrictions on forest management (use) (mean score: 4.52), and the form of compensation (mean score: 4.31). In contrast, the factors considered least determining included the impact on the local labor market (mean score: 3.07), the significance of the protection agreement (contract) at the national level (mean score: 3.11), and the initiator of the protection agreement (contract) (mean score: 3.31).

Table 4. Experts' perspectives on the factors influencing private forest owners' intentions to protect ecosystems.

Factors	Score of Likert Scale * Number of Experts/%					Mean Score	St. Deviation
	1	2	3	4	5		
Property rights and freedom of decision making	-	-	6/20.7	11/37.9	12/41.4	4.21	0.774
Amount of compensation	-	-	1/3.4	7/24.1	21/72.4	4.69	0.541
Clear definition of compensation	-	-	1/3.4	12/41.4	16/55.2	4.52	0.574
Cancelation policy for obligations	1/3.4	-	6/20.7	10/34.5	12/41.4	4.10	0.976
Compensation form	-	-	4/13.8	12/41.4	13/44.8	4.31	0.712
Duration of the protection agreement (contract)	-	2/6.9	2/6.9	17/58.6	8/27.6	4.07	0.799
Restrictions on forest management (use)	-	-	3/10.3	8/27.6	18/62.1	4.52	0.688
Continuity of the protection agreement (contract)	-	1/3.4	8/27.6	11/37.9	9/31.0	3.97	0.865
Distribution of compensation within the payment period specified in the protection agreement (contract)	-	1/3.4	7/24.1	14/48.3	7/24.1	3.93	0.799
Initiator of the protection agreement project (contract)	2/6.9	3/10.3	10/34.5	12/41.4	2/6.9	3.31	1.004
Achieving biodiversity and ecosystem protection goals	-	4/13.8	11/37.9	9/31.0	5/17.2	3.52	0.949
Impact on the local labor market	3/10.3	6/20.7	10/34.5	6/20.7	4/13.8	3.07	1.193
The significance of the protection agreement (contract) at the national level	1/3.4	8/27.6	9/31.0	7/24.1	3/10.3	3.11	1.066

* Value of score: 1—not determining; 2—slightly determining; 3—moderately determining; 4—highly determining; 5—extremely determining.

The responses to Question 9, "*What are the most important motives for the voluntary protection of biodiversity and ecosystems by Lithuanian private forest owners?*", are presented in Figure 7. The questionnaire categorized motives into three groups:

- *Internal motives*: these include the desire to improve environmental quality, enhance community well-being, preserve inheritance values, foster an emotional connection to the forest, engage in forest activities, and express or realize personal ideas in forest management.
- *Social motives*: these encompass factors such as reputation, image, moral satisfaction, and the desire to belong to or distinguish oneself from a group.
- *External motives*: primarily financial compensation.

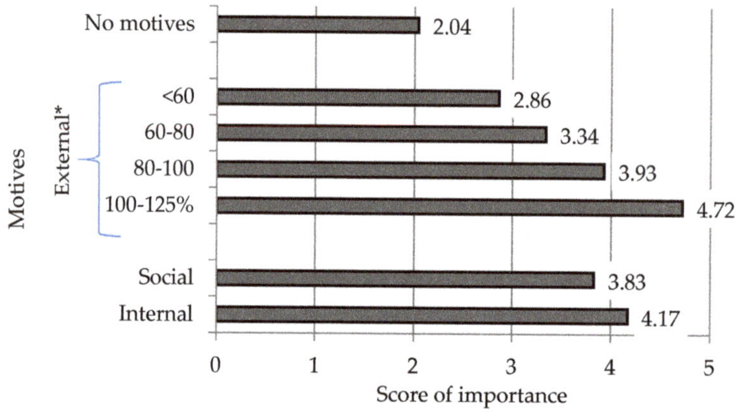

Figure 7. Experts' perspectives on the motivations for the voluntary protection of ecosystems by Lithuanian private forest owners. * External motive: monetary compensation % of incurred losses—100–125%, 80–100%, 60–80%, <60%.

According to the experts, external motives, particularly financial compensation amounting to 100–125% of incurred losses, were identified as the most-significant drivers for voluntary protection (mean score: 4.72). Internal motives were also rated as important (mean score: 4.17).

If compensation was less than 60% of the incurred losses, it would not sufficiently motivate private forest owners to voluntarily protect ecosystems. Therefore, when developing compensation schemes, it is important to ensure adequate coverage of losses, and low-intensity compensation schemes should be avoided.

The responses to Question 10, "*Should biodiversity and ecosystem protection be a moral obligation for private forest owners?*", are presented in Figure 8. The experts expressed divided opinions on this issue. Nearly half (44.8%) of the experts indicated that ecosystem protection should not be considered a moral obligation for private forest owners. Additionally, approximately 14% of the experts stated that they had no opinion on the matter.

The responses to Question 11, "*Would there be a "crowd-out" effect in Lithuania, whereby private forest owners with internal and social motivations might, if possible, opt for a Protection agreement (contract) with compensation?*", are presented in Figure 9. Nearly 38% of experts indicated that the "crowd-out" effect would not occur in Lithuania, while 31% expressed no opinion on the matter. These findings highlight the need for further research, particularly through direct surveys of private forest owners, to gain deeper insights into this issue.

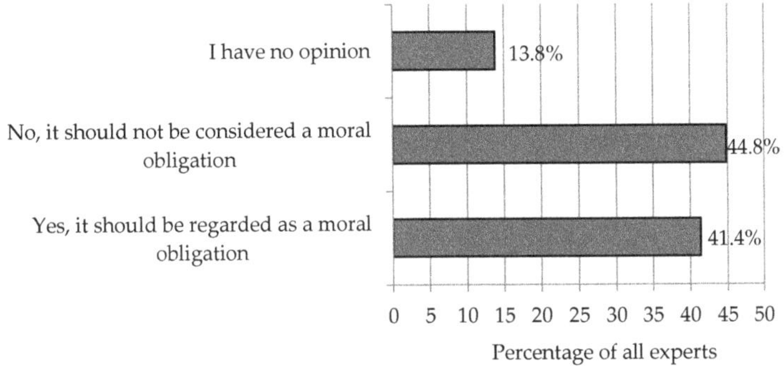

Figure 8. Experts' perspectives on the moral obligation of private forest owners to protect ecosystems.

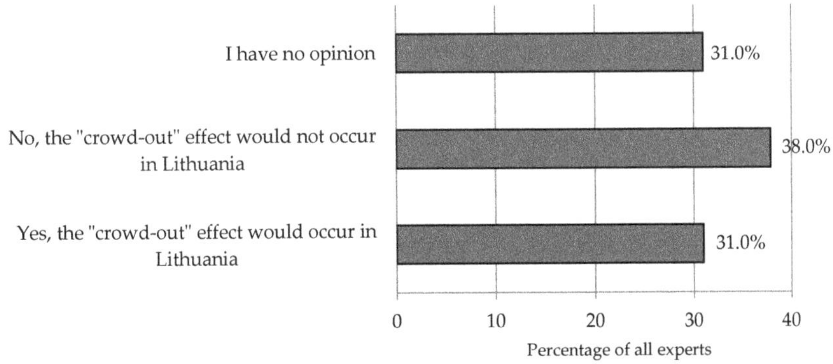

Figure 9. Experts' perspectives on the manifestation of the "crowd-out" effect in Lithuania.

The responses to Question 12, *"Do private forest owners incur losses in protecting biodiversity and ecosystems?"*, indicate that nearly 90% of experts suppose private forest owners incur losses when engaging in ecosystem protection. This underscores the need for a balanced and incentivizing compensation scheme in Lithuania, which would encourage voluntary participation by private forest owners in such protection efforts.

The responses to Question 13, *"Should losses incurred by private forest owners be compensated for biodiversity and ecosystem protection?"*, are presented in Figure 10. Nearly 97% of experts agreed that losses incurred by private forest owners should be compensated, with only 1 out of 30 experts opposing this view, arguing that compensation is unnecessary.

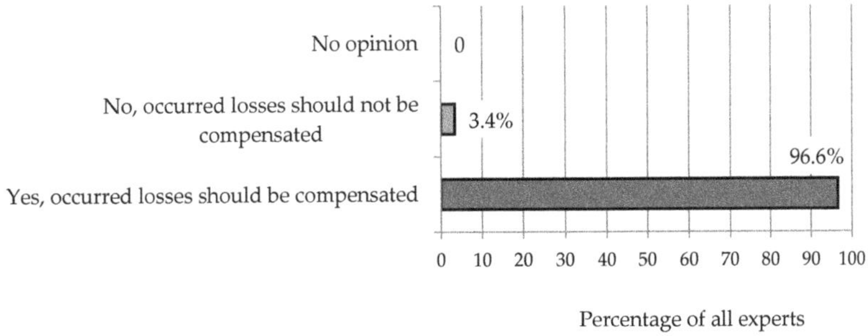

Figure 10. Experts' perspectives on the obligation to compensate private forest owners for losses incurred from ecosystem protection.

The responses to Question 14, *"What could be the sources of compensation for losses due to biodiversity and ecosystem protection?"*, are presented in Figure 11. Approximately 80% of experts identified the following potential sources of compensation: (1) EU funds allocated for the implementation of nature conservation programs (86.2%); (2) the Environmental Protection Support Program (79.3%); and (3) the Environmental Protection Support Special Program (SAARSP) (79.3%). Additionally, more than 50% of experts suggested the following sources could also be utilized: (1) state budget funding (72.4%); (2) foundations of private natural or legal entities declaring the implementation of environmental protection measures and the preservation of nature (65.5%); and (3) funds allocated for the optimization and redistribution of PAs (51.7%).

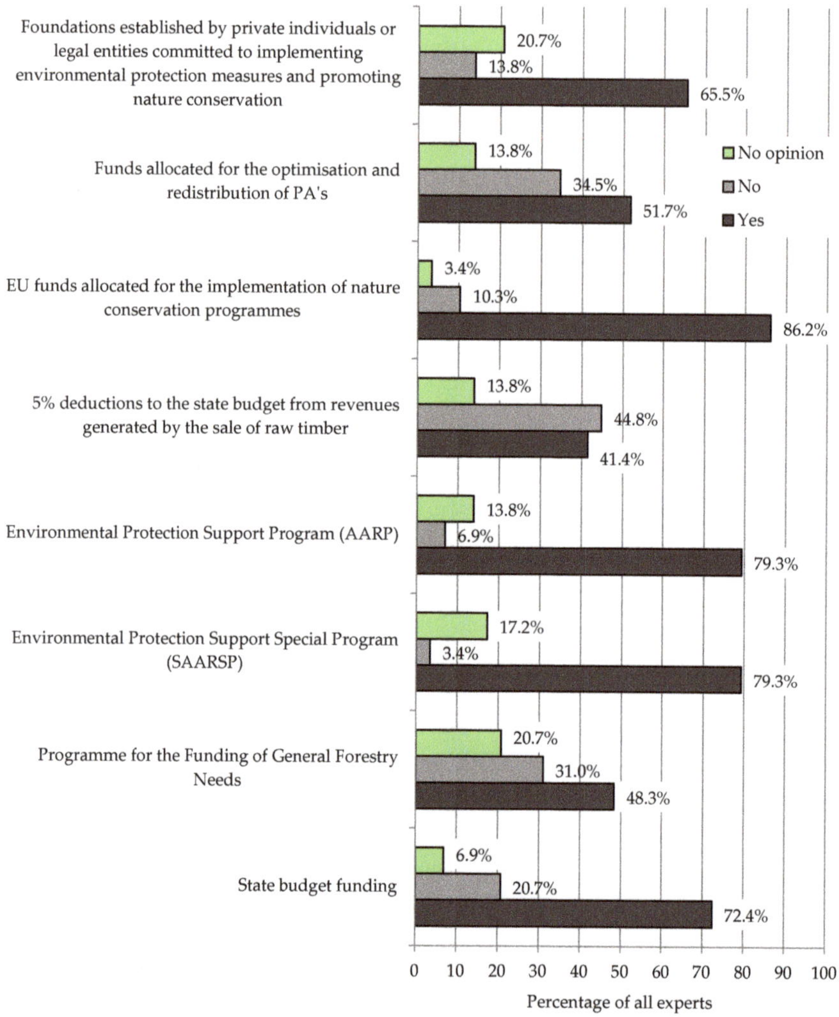

Figure 11. Experts' perspectives on potential sources of compensation for losses incurred in ecosystem protection.

The responses to Question 15, *"Are the currently applied measures for promoting biodiversity and ecosystem protection in private forests effective?"*, are presented in Figure 12. Among the three measures presented, experts identified the program *Support for Natura 2000 Forests* as the most effective; however, its effectiveness was rated at a relatively low mean score of 2.9. The measure receiving the lowest efficiency score was *Compensation for Activity Restrictions Imposed on Protected Areas Allocated Before the Restoration of Ownership Rights*, with a mean score of 2.38. Based on these evaluations, it can be concluded that none of the currently applied measures are considered effective.

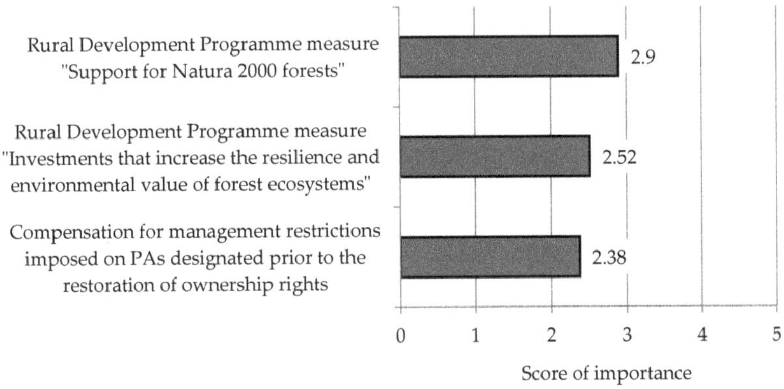

Figure 12. Assessment of the effectiveness of measures currently applied in Lithuania to promote ecosystem protection in private forests.

The responses to Question 16, "*If you suppose the current measures to promote biodiversity and ecosystem protection in private forests are ineffective, why?*", are presented in Figure 13. The experts identified the following key factors contributing to the ineffectiveness of these measures: (1) Payments (compensations) being unreasonable and excessively low (mean score = 4.28), (2) Payments (compensations) being insignificant in relation to the income of the forest owners (mean score = 4.24), (3) There being a lack of information for private forest owners (mean score = 4.03), and (4) Complex bureaucratic procedures for applying for compensations (mean score = 4.00).

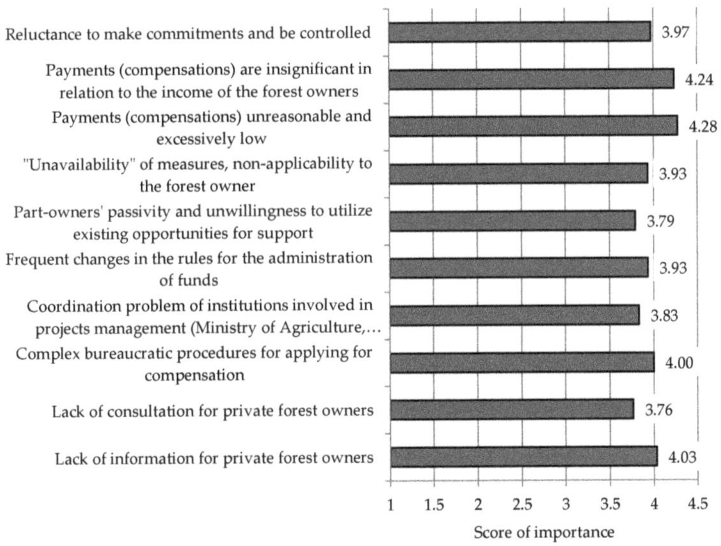

Figure 13. Factors and reasons contributing to the inefficiency of measures promoting ecosystem protection in private forests in Lithuania. Value of score: 1—not important; 2—slightly important; 3—moderately important; 4—very important; 5—extremely important.

4. Discussion

In this section, the priorities and motives of Lithuanian private forest owners regarding voluntary forest protection are analyzed from two perspectives: (1) in comparison with the results of similar studies conducted in other European countries, and (2) in relation to the private forest protection programs implemented in Lithuania.

Relevant studies from other European countries, referenced in this article and utilized for comparative analysis with the Lithuanian context, highlight both the specific and general aspects of Lithuania's situation. The study by Kosenius [3] identified the heterogeneity of private forest owners' perspectives on voluntary forest protection, demonstrating its dependence on various characteristics such as forest land size, gender, and place of residence. This heterogeneity was also evident in Lithuania, where unambiguous responses were absent.

Lithuanian experts showed a stronger preference for a permanent protection model with compensation while maintaining ownership (72.4%) compared to contracts with a fixed duration of 10–20 years (48.3%). In contrast, Finnish private forest owners generally favor shorter-term contracts [3]. When addressing the challenges of voluntary forest protection in Lithuania and other countries, significant attention is given to compensating for losses resulting from restrictions on economic activity. In analyzing the perspectives of Latvian private forest owners, Baranovskis et al. concluded that income reduction is the primary factor contributing to negative attitudes towards forest protection. Similarly, a study by Kosenius in Finland [3] emphasized the necessity of providing higher compensation to offset restrictions on economic activities. Finnish experiences suggest that voluntary forest protection schemes would be more attractive to private forest owners if compensation exceeded the financial losses incurred [14,15]. Palome [10] found that economic incentives play a significant role in encouraging participation in biodiversity-related protection programs. Comparable conclusions can be drawn from Lithuanian expert responses, which identified the key factors influencing private forest owners' willingness to engage in ecosystem protection as the compensation amount (Likert scale score: 4.69), clear definition of compensation (4.52), and restrictions on forest management (4.52). Lithuanian private forest owners are mostly motivated by financial compensation of 100–125% of losses. They do not support forest protection models without compensation (96.6% of experts). In Norway, only 3% of private forest owners support non-compensation models [4].

The amount of compensation is typically linked to losses incurred in timber use. In Lithuania, compensation may be provided when forest felling is prohibited. The Law on Protected Areas of the Republic of Lithuania stipulates that compensation is granted for the income that could have been obtained from selling timber on the market, minus the average logging costs. Compensation is calculated based on the average market prices of timber and the average logging costs in the year in which the private forest owner submits a request for its calculation and payment.

An expert survey indicated that the mechanisms for compensating losses due to the establishment of PAs require improvement. According to Lithuanian experts, two protection models were considered the most appropriate: (1) permanent protection with compensation while retaining forest ownership (72.4%) and (2) selling forests to the state (55.2%). In Finland [6], most respondents (27%) expressed a preference for fixed-term protection contracts, while 18% favored permanent protection contracts, and 17% were inclined to sell their forests to the state for protection. In Lithuania, selling forests designated for protection to the state is more widely accepted, with 55.2% of experts considering this model viable. The opinions of Lithuanian and Norwegian experts differ regarding the proportion of PAs. The current share of protected areas is considered optimal by 17.2% of Lithuanian experts and 47.0% of Norwegian experts [4]. However, 13.8% of Lithuanian experts believe that the extent of protected areas is insufficient, compared to only 3.0% in Norway. Conversely, 55.2% of Lithuanian experts and 20.0% of Norwegian experts consider the proportion of protected areas excessive. The views of Swedish private forest owners align more closely with those of Norwegian experts [8], with 73.4% of respondents stating that the current extent of protected areas is sufficient or too much, 17.3% believing it exceeds the necessary level, and 10.0% advocating for increased protection.

The opinion of Lithuanian experts aligns with the findings of Juutinen et al. [7], which highlight the heterogeneity of views among forest owners, the high compensation required for contract-based forest management, and the necessity of additional information and education.

Regarding the "crowding-out" effect, which has been identified in studies examining the motivations of French forest owners to engage in forest protection [10], Lithuanian experts expressed divided opinions: 31% had no opinion, 38% believed that the "crowding-out" effect would not occur in Lithuania, while 31% considered that it would. Institutional factors play a significant role in forest owners' commitments to biodiversity protection [9]. However, Lithuanian experts assigned a relatively low importance to the factor "Initiator of the protection agreement (contract)", rating it at 3.31 on a five-point Likert scale.

Findings from studies conducted in other countries suggest that a positive environmental attitude among forest owners increases the likelihood of participation in forest protection programs [11]. Additionally, enhancing knowledge of economic, managerial, and attitudinal factors significantly influences ecosystem protection processes [12]. Furthermore, the necessity of conflict analysis between private forest owners and authorities [13] is an important consideration for developing voluntary private forest protection schemes in Lithuania.

Both international studies and this research highlight that voluntary private forest protection in Lithuania has been primarily examined through the lens of financial compensation, potentially overlooking non-economic values associated with biodiversity conservation, as well as other social and cultural drivers. To improve the development of voluntary private forest protection, future research should incorporate questionnaire items that explore these non-economic values and sociocultural motivations underlying forest conservation efforts.

Lithuanian experts often express a negative view of the current private forest protection regime. The effectiveness of existing programs was evaluated using a five-point rating scale, where 1 indicates complete ineffectiveness and 5 represents high effectiveness. The most effective measure, according to experts, was the Rural Development Programme measure "Support for Natura 2000 forests", yet it was rated at only 2.9. The least-effective measure, the Rural Development Programme measure "Compensation for management restrictions imposed on protected areas designated prior to the restoration of ownership rights", received an even lower mean score of 2.38. Lithuanian experts point out the most-important factors and reasons for such inefficiency: a large part of losses due to restrictions in PAs are not compensated at all, compensation is insignificant concerning the income of the forest owners, compensation is unreasonable and excessively low, there is insufficient information given to private forest owners, and there are complex bureaucratic procedures associated with applying for support. Studies [31] and the media [32,33] have found that compensation does not match actual losses.

5. Conclusions

The forest protection priorities of Lithuanian private forest owners regarding the development of PAs in private forests, protection models, and protection promotion motives often coincide with the priorities and motives established in other European countries. A heterogeneity of priorities and motives was identified. Almost three-quarters of experts thought the current amount of protected forest in Lithuania is sufficient or is already more than necessary, and only about one in ten thought that it is necessary for owners to protect more forest. Lithuanian private forest owners are mostly motivated by full financial compensation for losses. They do not support forest protection models without compensation. It would be appropriate to apply both permanent and fixed-term protection agreements (contracts) with compensations, as well as the option of selling forests to the state. The amount of compensation is the most-significant factor influencing private forest owners' willingness to protect ecosystems. The primary reasons for the ineffective private forest protection measures in Lithuania include inadequate and unjustified compensations, compensations too insignificant for forest owners' income, a lack of information, and complex bureaucratic procedures.

This study's findings are pertinent to the State Service for Protected Areas under the Ministry of Environment of Lithuania, which oversees the development of protected areas in the country. Additionally, private forest owners and their associations, who have a vested interest in voluntary ecosystem protection, may find the insights valuable. As the inaugural study on voluntary ecosystem protection in Lithuania, it underscores the necessity of addressing critical issues such as forest suitability for protection and compensation mechanisms for economic restrictions imposed on forest owners. Future research should focus on these areas to enhance the effectiveness and acceptance of voluntary protection initiatives.

Author Contributions: Conceptualization, D.L. and S.M.; methodology, D.L.; data analysis, D.L.; writing—original draft preparation, D.L.; writing—review and editing, S.M. All authors have read and agreed to the published version of the manuscript.

Funding: This research was funded by the Ministry of Environment of Republic of Lithuania, agreement number VPS-2022-5-SBMŪRP.

Data Availability Statement: The raw data supporting the conclusions of this article will be made available by the authors on request.

Conflicts of Interest: The authors declare no conflicts of interest. The funders had no role in the design of this study; in the collection, analyses, or interpretation of data; in the writing of the manuscript; or in the decision to publish the results.

References

1. Civic, K. Voluntary Protection of Forests in Finland, Sweden and Norway. European Land Conservation Network. 2018. Available online: https://elcn.eu/index.php/news/voluntary-protection-forests-finland-sweden-and-norway.html (accessed on 16 April 2024).
2. Frank, G.; Muller, F. Voluntary approaches in the protection of forests in Austria. *Environ. Sci. Policy* **2006**, *6*, 261–269. [CrossRef]
3. Kosenius, A.K. Forest Owner Attitudes and Preferences for Voluntary Temporary Forest Conservation. *Small-Scale For.* **2024**, *23*, 493–513. [CrossRef]
4. Gundersen, V.; Vistad, O.I.; Skjeggedal, T. Forest owners' perspectives on forest protection in Norway. *Scand. J. For. Res.* **2022**, *37*, 276–286. [CrossRef]
5. Baranovskis, G.; Nikodemus, O.; Brūmelis, G.; Elferts, D. Biodiversity conservation in private forests: Factors driving landowners attitude. *Biol. Conserv.* **2022**, *266*, 109441. [CrossRef]
6. Koskela, T.; Karppinen, H. Forest Owners Willingness to Implement Measures to Safeguard Biodiversity: Values, Attitudes, Ecological Worldview and Forest Ownership Objectives. *Small-Scale For.* **2021**, *20*, 11–37. [CrossRef]
7. Juutinen, A.; Kurttila, M.; Pohjanmies, T.; Tolvanen, A.; Kuhlmey, K.; Skudnik, M.; Triplat, M.; Westin, K.; Makipaa, R. Forest owner's preferences for contract-based management to enhance environmental values versus timber production. *For. Policy Econ.* **2021**, *132*, 102587. [CrossRef]
8. Danley, B.; Bjärstig, T.; Sandström, C. At the limit of volunteerism? Swedish family forest owners and two policy strategies to increase forest biodiversity. *Land Use* **2021**, *105*, 105–403. [CrossRef]
9. Abildtrup, J.; Stenger, A.; de Morogues, F.; Polome, P.; Blondet, M.; Michel, C. Biodiversity Protection in Private Forests: PES Schemes, Institutions and Prosocial Behavior. *Forests* **2021**, *12*, 1241. [CrossRef]

10. Polome, P. Private forest owners' motivation for adopting biodiversity-related protection programs. *J. Environ. Manag.* **2016**, *183*, 212–219. [CrossRef] [PubMed]
11. Mitani, Y.; Lindhjem, H. Forest Owners' Participation in Voluntary Biodiversity Conservation: What Does it Take to Forgo Forestry for Eternity? *Land Econ.* **2015**, *91*, 235–251. [CrossRef]
12. Górriz, E.; Mäntymaa, E.; Petucco, C.; Schubert, F.; Vedel, S.E.; Mantau, U.; Prokofieva, I. Explaining participation of private forest owners in economic incentives. Case studies in Europe. In Proceedings of the Scandinavian Forest Economics: Proceedings of the Biennial Meeting of the Scandinavian Society of Forest Economics, Uppsala, Sweden, 21–24 May 2014; Volume 45, pp. 80–89. Available online: https://efi.int/sites/default/files/files/efimed/resources/Explaining%20participation%20of%20private%20forest%20owners%20in%C2%A0economic%20incentives.%20Case%20studies%20in%20Europe.pdf (accessed on 16 April 2024).
13. Götmark, F. Conflicts in conservation: Woodland key habitats, authorities and private forest owners in Sweden. *Scand. J. For. Res.* **2009**, *24*, 504–514. [CrossRef]
14. METSO. Voluntary Forest Protection Popular Among Forest Owners—Record Numbers of Environmental Aid Agreements and Nature Management Measures Under METSO Programme. *METSO News*, 14 February 2024. Available online: https://valtioneuvosto.fi/en/-/1410903/voluntary-forest-protection-popular-among-forest-owners-record-numbers-of-environmental-aid-agreements-and-nature-management-measures-under-metso-programme (accessed on 18 September 2024).
15. METSO. Almost 90 Percent of the Protection Target Set in the METSO Programme has Already Been Reached—More Funding is Needed for Environmental Subsidies and Nature Management. *METSO News*, 17 July 2022. Available online: https://metsonpolku.fi/en/-/almost-90-per-cent-of-protection-target-set-in-metso-programme-already-reached-more-funding-needed-for-environmental-subsidies-and-nature-management (accessed on 16 April 2024).
16. Syrjänen, K.; Apala, K.; Anttila, S.; Kuusela, S. Success and Challenges of Voluntary Forest Conservation in Finland. In Proceedings of the 5th European Congress of Conservation Biology, Jyväskylä, Finland, 12–15 June 2018; Available online: https://doi.org/10.17011/conference/eccb2018/107741 (accessed on 18 September 2024).
17. Ymparisto.fi. METSO—The Forest Biodiversity Programme for Southern Finland. *Finnish Environment Institute (Syke)*, 19 June 2023. Available online: https://www.ymparisto.fi/en/nature-waters-and-seas/natural-diversity/conservation-and-research-programmes/metso-forest-biodiversity-programme-southern-finland (accessed on 16 April 2024).
18. Borg, R.; Paloniemi, R. Deliberation in cooperative networks for forest conservation. *J. Integr. Environ. Sci.* **2012**, *9*, 151–166. [CrossRef]
19. Mäntymaa, E.; Juutinen, A.; Mönkkönen, M.; Svento, R. Participation and compensation claims in voluntary forest conservation: A case of privately owned forests in Finland. *For. Policy Econ.* **2009**, *11*, 498–507. [CrossRef]
20. METSO. *Government Resolution on the Forest Biodiversity Programme for Southern Finland 2008–2016 (METSO-Programme)*; METSO: Helsinki, Finland, 2008.
21. Hiedanpää, J. The edges of conflict and consensus: A case for creativity in regional forest policy in Southwest Finland. *Ecol. Econ.* **2005**, *55*, 485–498. [CrossRef]
22. Horne, P.; Ovaskainen, V.; Koskela, T. Economic and Social Implications of Incentive-Based Policy Mechanisms in Biodiversity Conservation. In *Working Papers of Finnish Forest Research Institute*; Finnish Forest Research Institute: Helsinki, Finland, 2004; p. 168. Available online: https://jukuri.luke.fi/bitstream/handle/10024/535918/mwp001.pdf?sequence=1 (accessed on 18 September 2024).
23. Thomasson, T. *Payments for Ecosystem Services: What role for a green economy?* Swedish Forest Agency: Jönköping, Sweden, 2011; p. 10. Available online: https://unece.org/fileadmin/DAM/timber/meetings/20110704/2-KOMET_Sweden_Tove_Thomasson_Geneva_5_July_2011.pdf (accessed on 18 September 2024).
24. Naturvårdsverket. Kometprogrammet 2010–2014. In *Naturvardsverket Raport 2014, 6621*; Naturvårdsverket: Stockholm, Sweden, 2014; p. 57. Available online: https://naturvardsverket.diva-portal.org/smash/get/diva2:1389402/FULLTEXT01.pdf (accessed on 18 September 2024).
25. Widman, U. Exploring the Role of Public-Private Partnership in Forest Protection. *Sustainability* **2016**, *8*, 496. [CrossRef]
26. Aaro, I. Voluntary Protection of Forests in Norway. CEPF Blank Page. 2022. Available online: https://www.facebook.com/photo?fbid=6405969259418430&set=a.207454412603310 (accessed on 16 April 2024).
27. Dalkey, N. *The Delphi Method: An Experimental Study of Group Opinion*; Rand: Santa Monica, CA, USA, 1969; p. 90. Available online: https://www.rand.org/content/dam/rand/pubs/research_memoranda/2005/RM5888.pdf (accessed on 18 September 2024).
28. Dalkey, N. An experimental study of group opinion. *Futures* **1969**, *1*, 408–426. [CrossRef]
29. Colton, D.; Covert, R.W. *Designing and Constructing Instruments for Social Research and Evaluation*; Jossey-Bass: Honiken, NJ, USA, 2007; p. 394. Available online: https://www.wiley.com/en-in/Designing+and+Constructing+Instruments+for+Social+Research+and+Evaluation-p-9780787987848#download-product-flyer (accessed on 16 April 2024).

30. Aiken, L. *Attitudes and Related Psychosocial Constructs: Theories, Assessments, and Research;* Sage Publications: Thousand Oaks, CA, USA, 2002; p. 328. Available online: https://sk.sagepub.com/book/mono/attitudes-and-related-psychosocial-constructs/toc (accessed on 18 September 2024).
31. Brukas, V.; Stanislovaitis, A.; Kavaliauskas, M.; Gaižutis, A. Protecting or destructing? Local perceptions of environmental consideration in Lithuanian forestry. *Land Use Policy* **2018**, *79*, 1014–1023. [CrossRef]
32. Tavorienė, V. ES Svarbos Buveinės. Tikėjosi Grąžos, bet Gavo Šnipštą. 2022. Available online: https://www.forest.lt/Naujienos/es-svarbos-buveines-tikejosi-grazos-bet-gavo-snipsta/ (accessed on 16 April 2024).
33. Mano ūkis. Kaip Valstybei Pavyks Susitarti su Savininkais dėl ST Plėtros? *Mano ūkis*, 25 February 2021. Available online: https://www.manoukis.lt/naujienos/aplinka-miskai/kaip-valstybei-pavyks-susitarti-su-savininkais-del-saugomu-teritoriju-pletros (accessed on 16 April 2024).

land

MDPI

Article

Citizen Science to Investigate the Ecophysiological Responses of Mediterranean Shrubland Vegetation in an Urban Open-Air Laboratory

Valerio Lazzeri [1], Francesca Bretzel [1,2], Graziella Rossini [3,4], Laura Pellegrino [3,4], Laura Marchetti [5], Rosella Panti [5], Enrica Talà [5], Antonio Martelli [4] and Andrea Scartazza [1,2,*]

1 Research Institute on Terrestrial Ecosystems (IRET), National Research Council (CNR), Via Moruzzi 1, 56124 Pisa, Italy; valerio.lazzeri@iret.cnr.it (V.L.); francesca.bretzel@cnr.it (F.B.)
2 National Biodiversity Future Center (NBFC), 90133 Palermo, Italy
3 Institute of Bioeconomy (IBE), National Research Council (CNR), Viale Italia 6, 57126 Livorno, Italy; graziella.rossini@ibe.cnr.it (G.R.); laura.pellegrino@ibe.cnr.it (L.P.)
4 Association World Wide Fund for Nature (WWF) Livorno, Via Don Bosco 7, 57124 Livorno, Italy; antonio.martelli@tim.it
5 Istituto Comprensivo G. Bartolena, Plesso di Villa Corridi, Via del Vecchio Lazzeretto 10, 57124 Livorno, Italy; marchetti.l@scuolabartolena.edu.it (L.M.); panti.r@scuolabartolena.edu.it (R.P.); talarica1961@gmail.com (E.T.)
* Correspondence: andrea.scartazza@cnr.it

Abstract: Urban biodiversity is essential for enhancing liveability for both humans and wildlife by providing a range of ecosystem services. Therefore, it is crucial to raise awareness among citizens, particularly schoolchildren, about the significance of biodiversity in urban environments. To this end, an Open-Air Laboratory was established in an urban park of an Italian Mediterranean city (Livorno), where natural vegetation flourishes. This initiative engaged schoolchildren in experimental campaigns to collect ecophysiological data on local wild woody species. Specifically, the students were tasked with identifying various wild woody species and, under the guidance of researchers, recording specific leaf ecophysiological traits, such as leaf mass per area, leaf pigments, and chlorophyll fluorescence. The results, which highlighted seasonal variations in leaf ecophysiological traits and interspecific differences, were analysed in relation to the environmental conditions documented by the schoolchildren. This analysis revealed distinct plant strategies for coping with winter and summer stressful periods. The methodology employed in this project, which involved schoolchildren in research activities, not only fostered environmental awareness among young participants but also serves as a pilot model for public engagement in scientific research.

Keywords: ecosystem services; informal green spaces; Mediterranean maquis; photosynthesis; plant biodiversity; schoolchildren biodiversity education

1. Introduction

Plant diversity plays a significant role in Mediterranean urban and suburban landscapes. Nowadays, it is well established that plant communities provide a multitude of ecosystem services that contribute to the overall liveability of urban environments [1,2]. In this context, it has been shown that the more biodiverse these communities are, the greater their contribution to ecosystem services [3,4]. Among these services, woody species can mitigate the urban heat island effect during the summer, improve air quality by removing particulate matter and gaseous pollutants, and enhance the aesthetic value of urban areas, as well as the restorative potential of green spaces [5,6]. Furthermore, many urban areas, including those recently developed, lack green spaces, resulting in environments that are unsuitable for wild flora and fauna, which contributes to an unhealthy environment for humans [7]. In this regard, urban parks, due to their proximity to city centres and ease of access, can be viewed as Open-Air Laboratories (OALs) where biodiversity and plant ecophysiology can be studied, with the assistance of visitors participating in citizen science activities [8,9]. Urban green spaces that maintain natural or semi-natural surfaces are particularly ideal for conducting such studies [10,11]. These areas are also referred to as "informal green spaces" and are recognised as providers of numerous ecosystem services, including the facilitation of educational activities [12].

Raising public awareness of environmental issues and the importance of biodiversity conservation among citizens is crucial, especially for children, who are tomorrow's adults. Some authors have emphasised the significance of concepts such as the "extinction of experience" and the excess of virtual experiences that focus primarily on exotic species [13,14]. Engaging individuals, particularly schoolchildren, in hands-on, environment-related scientific research can be essential for making these topics more accessible and fostering a desire to learn more [15]. From this perspective, involving them in activities in OALs is an effective way to enhance their learning experience [16,17]. Citizen science is regarded as a powerful means of implementing initiatives aimed at biodiversity conservation [18] and has the potential to contribute to the United Nations' Sustainable Development Goals [19], as well as to help reduce inequality [20]. Furthermore, identifying locations that are both easily accessible and sufficiently safe for schoolchildren, along with providing plant species suitable for an OAL, is a significant milestone in enhancing the general public's environmental awareness and knowledge. Cities, particularly urban natural and semi-natural areas, can serve as laboratories for studying and evaluating the effectiveness of Nature-based Solutions (NbS), which seek to utilise native wild plant species for urban green infrastructures [21].

The Mediterranean urban environment is of particular interest for several reasons. Plant species that grow at low altitudes, especially during the dry season, must be able to tolerate severely stressful conditions, such as high temperatures and water scarcity [22]. The adaptation of the photosynthetic apparatus to Mediterranean environmental conditions results in these ecosystems exhibiting some of the highest net primary productivities worldwide [23]. Consequently, Mediterranean species are particularly intriguing due to their photosynthetic strategies before, during, and after the most stressful periods. Their ability to cope with such stresses makes them a viable choice for cultivation in urban green areas [24]. Thanks to their greater plasticity and variability in functional traits, they can survive in extreme environmental conditions, thereby enhancing overall ecosystem resilience and providing maximum ecosystem services [25]. The primary school curriculum in Italy encompasses the study of living organisms, including their structures, functions, and interactions with the environment. While photosynthesis is typically addressed in lower secondary school, previous research indicates that primary students can grasp this process to a certain extent [26]. Furthermore, the direct involvement of schoolchildren in studying seasonal changes in plants is particularly beneficial, especially when scientific instruments are utilised [27]. However, it is important to note the lack of projects that engage citizens in studying ecosystem functions [28], particularly those that provide young students with direct involvement in measuring photosynthesis.

To enhance public awareness of issues related to urban plant biodiversity and its interaction with the environment, an OAL has been established in the urban park of Villa Corridi, located in the Mediterranean coastal city of Livorno, Tuscany, Italy. This initiative leverages the presence of primary and secondary schools within the park, as well as the naturally occurring Mediterranean maquis vegetation. The OAL serves as a living laboratory for schoolchildren, aiming to (i) introduce them to the scientific study of plant biodiversity in Mediterranean thermophile woodlands; and (ii) encourage students to explore these environments and measure ecophysiological leaf traits using scientific instruments and methods employed by researchers to generate replicable and reliable scientific data.

2. Materials and Methods

2.1. Site Description

Livorno is a city situated along the coast of Tuscany (northern Central Italy), with a population of approximately 153,000 residents and an area of about 104 km^2. The chemical industry and harbour activities play a key role for the city's economy and contribute significantly to its overall pollution status. The climate of Livorno is, according to the Köppen–Geiger Climate Classification, typically Mediterranean climate, characterised as warm temperate with dry, hot summers (Csa) [29,30] (Figure S1). In terms of bioclimatic classification [31], Livorno exhibits a Mediterranean macrobioclimate, a transitional Oceanic–Mediterranean bioclimate, a Mesomediterranean thermotype, and a subhumid ombrotype. The Villa Corridi park is situated on the outskirts of Livorno (Figure 1a) and spans approximately 114,000 m^2. The park is primarily designed as a formal green space, encompassing around 81,200 m^2 of lawns and trees, mainly Aleppo pines. However, over the past 50 years, a semi-natural vegetation has developed on its south-facing side. This area features cultivated pines and trails that facilitate access for visitors while still showcasing species typical of the Mediterranean maquis, which is well-represented in protected areas of the surrounding hillside.

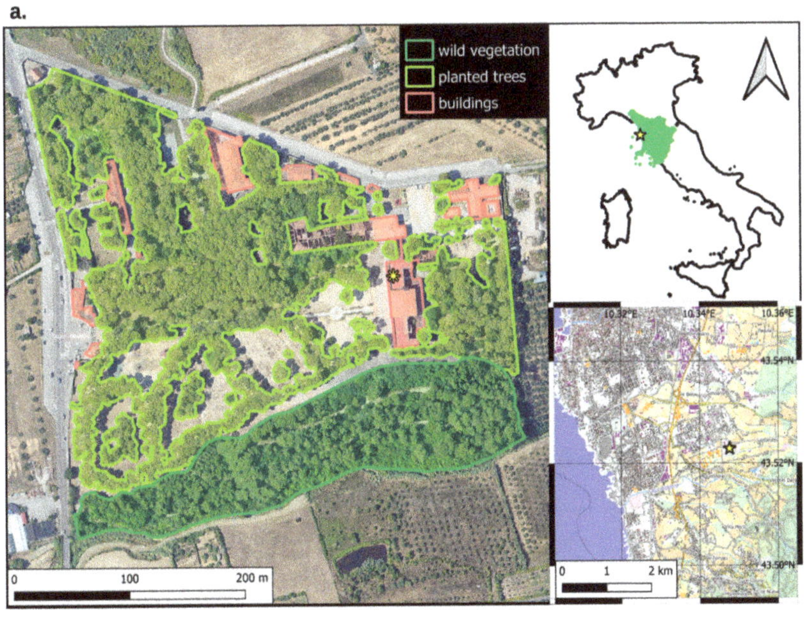

Figure 1. *Cont.*

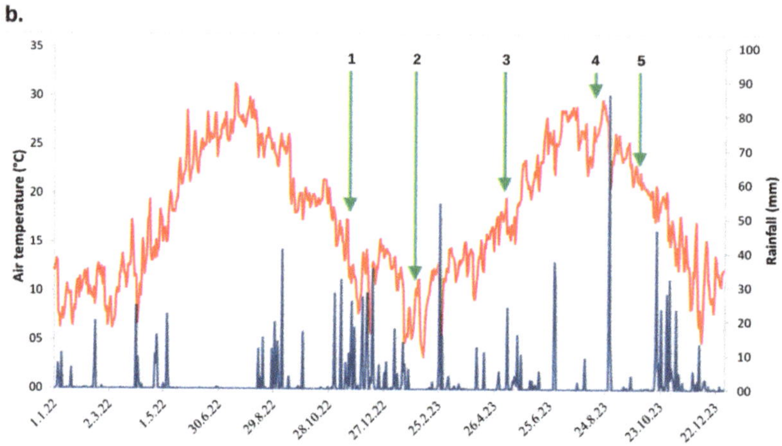

Figure 1. Site vegetation and thermopluviometric chart: (**a**) Left: cartography of the vegetation in the study site (created with QGIS 3.22; 43°31.373′ N 10°20.899′ E). Deep green, wild vegetation (study area); light green: planted trees; pink, buildings; asterisk, position of the meteorological station. Cartographic basis: Google Earth. Bottom right: star, position of the study site in relation to the city of Livorno. Cartographic basis: Regione Toscana. Top right: star, position of the study site in Tuscany. Cartographic basis: Regione Toscana. (**b**) Daily average air temperatures and rainfall measured during the two years of investigation. Green arrows indicate the sampling dates.

2.2. Classroom Activities and the Establishment of an OAL with Students and Teachers

Two classes from the 5th grade of primary school (10 years old) and the 2nd grade of lower secondary school (11–12 years old), along with their teachers, participated in the implementation of an OAL in the park of Villa Corridi during two events held in Autumn 2022 and Spring 2023. A preparatory session was conducted in the classroom, covering topics such as biodiversity on Earth and plant ecophysiology, which were aligned with the objectives of the activities carried out in the park. In this context, a brief introduction to the citizen science application iNaturalist was also provided. Care was taken to employ a "teaching through questioning" approach [32], while avoiding an excessively frontal method. Following the field activities, additional classroom meetings were held to present the results to both the schoolchildren and teachers. These sessions utilised a brainstorming approach, allowing participants to express their ideas spontaneously and fostering creative thinking. Laboratory sessions focused on meteorology and climatology were conducted for schoolchildren and their teachers. These sessions included specific lessons on the operation of the meteorological station, the process of acquiring various meteorological parameters, and the recording of archived data, along with their applications in different fields. To broaden the OAL experience to a wider audience, additional training sessions were organised in the park of Villa Corridi between 2022 and 2023, engaging other schoolchildren and their teachers. These sessions included the same activities conducted in the OAL and focused on plant species identification, the evaluation of their photosynthetic performance, and the collection of meteorological data.

2.3. Acquisition of Meteorological Data

The environmental conditions of the Villa Corridi park were recorded by schoolchildren and their teachers using a Davis ISS Vantage Pro 2 Plus meteorological station (Davis Instruments, Hayward, CA, USA) equipped with various sensors, including a rain gauge, thermo-hygrometer, cup anemometer, solar radiation sensor, UV radiation sensor, and a Vantage Pro 2 Plus Console unit with a barometer and internal thermo-hygrometer

(Figure 1a). The station has been installed since 2017 on the terrace of Villa Corridi Primary School as part of an agreement ratified between CNR IBE and the school board for educational and research activities outlined in the "ECOPLANET" macro-project, specifically the "Green Action Research" sub-project. This initiative aims to enhance understanding of climate change and implement monitoring activities to support outdoor education in pursuit of the Sustainable Environmental Goals of the 2030 Agenda. From January 2022 to December 2023, students and teachers collected and analysed data, including air temperature and daily mean precipitation (Figure 1b), to evaluate the effects of seasonal environmental conditions on the monitored leaf functional traits.

2.4. *Survey of Plant Diversity in the Urban Park*

A survey of the wild vegetation was conducted to identify the shrub and tree species most prevalent in the park, which, with the exception of one, are characteristic of the sclerophyllous Mediterranean maquis and woodlands in the surrounding areas. Among the observed species, a list was compiled that includes those taxa selected for ecophysiological analysis: *Arbutus unedo* L., *Cistus creticus* L. subsp. *eriocephalus* (Viv.) Greuter & Burdet, *Laurus nobilis* L., *Ligustrum lucidum* W.T.Aiton, *Myrtus communis* L. subsp. *communis*, *Pinus halepensis* Miller, *Phillyrea angustifolia* L., *P. latifolia* L., *Pistacia lentiscus* L., *Quercus ilex* L., *Rhamnus alaternus* L. subsp. *alaternus*, and *Viburnum tinus* L. subsp. *tinus*. Subsequently, a subset of eight taxa was selected for activities with schoolchildren to align with the school schedule: *C. creticus* subsp. *eriocephalus*, *L. nobilis*, *M. communis*, *P. halepensis*, *P. lentiscus*, *Q. ilex*, *R. alaternus*, and *V. tinus* subsp. *tinus* (Figures S2–S4). These species were chosen to provide a set of diagnostic morphological characters that are (a) easy for beginners to evaluate, (b) representative of the most important characters used in botany, and (c) involve not only the sense of sight but also touch and smell (Table 1). Specifically, information cards were prepared for each species, featuring a colour photograph of the plant sourced from non-copyrighted botanical works, along with a brief description and a simple list of diagnostic characters (the English version in the Supplementary Material, Figures S2 and S3). It is important to note that in these cards, the terms "shrub" and "tree" are used not in a strict botanical sense, but rather to distinguish between plants that are up to two meters tall and those that exceed this height. Subsequently, schoolchildren were divided into groups, with each group receiving two cards at a time. The selection of the two species presented simultaneously was designed to encourage pupils to compare two types of plants that clearly differed in several morphological characters. The students were then encouraged to collaborate in order to identify a small specimen of the correct species. A collaborative approach was promoted to minimise competition between groups, while emphasizing that the project's goals were to enhance their knowledge of wild flora and its ecology, to collaborate with researchers, and to have fun. Additionally, the children's observational skills, along with their ability to connect a simple text and an image to the living plants, were carefully nurtured. Efforts were also made to structure the activity as a game, making the experience more engaging for the pupils and maintaining their attention [33]. Once the specimens were collected, they were verified by experts. In cases of misidentification, the specific errors were explained. No scores were assigned for correct identifications, nor was a ranking established. Teachers were then asked to photograph the identified species and upload the georeferenced images to the citizen science application iNaturalist to map the observed plants.

Table 1. Main morphological diagnostic features of the studied species.

Species	Habitus	Leaves						
		Phyllotaxy	Simple (s) or Compound (c)	Colour	Texture	Indumentum	Margin	Smell
Arbutus unedo	tree	alternate	s	green, slightly discolorous, darker on adaxial side	leathery	glabrous	dentate	grassy
Cistus creticus	shrub	opposite	s	greyish-green, concolorous	rather soft	hairy	entire	grassy
Laurus nobilis	tree	alternate	s	dark green, concolorous	leathery	glabrous	almost entire, usually wavy	pleasant
Myrtus communis	shrub	opposite	s	light green, concolorous	leathery	glabrous	entire	pleasant
Pinus halepensis	tree	fascicled	s	light green	-	glabrous	-	pleasant
Phillyrea angustifolia	shrub	opposite	s	green, slightly discolorous, darker on adaxial side	leathery	glabrous	usually entire, rarely dentate	grassy
Phillyrea latifolia	shrub	opposite	s	green, slightly discolorous, darker on adaxial side	leathery	glabrous	usually dentate, rarely entire	grassy
Pistacia lentiscus	shrub	alternate	c	green, slightly discolorous, darker on adaxial side	leathery	glabrous	entire	pleasant or un-pleasant
Quercus ilex	tree	alternate	s	strongly discolorous, dark green on adaxial side, whitish on abaxial side	leathery	glabrous on adaxial side, densely hairy on abaxial side	usually entire, rarely dentate	grassy
Rhamnus alaternus	shrub	alternate	s	green, slightly discolorous, darker on adaxial side	leathery	usually glabrous	dentate, or entire	foetid
Viburnum tinus	shrub	opposite	s	green, slightly discolorous, darker on adaxial side	rather soft	hairy especially on margins	entire	grassy

2.5. Determination of Leaf Ecophysiological Traits in Mediterranean Vegetation

To promote understanding of functional biodiversity, the schoolchildren participated in experimental field campaigns alongside scientists to record leaf ecophysiological traits, such as leaf mass per area (LMA) and photosynthetic performance (Figure S4). Sampling dates and the number of replicates were determined while considering the constraints imposed by the school calendar and timetable. Consequently, to facilitate seasonal comparisons, the same number of replicates per species was maintained during the other sampling dates as well.

For the determination of LMA, leaf disks (1.3 cm^2) and needles (for *P. halepensis*) of a known area were collected, placed in pre-weighed vials, and transported to the laboratory. In the lab, the leaf disks were oven-dried at 50 °C until they reached a constant weight, and then weighed to determine the dry weight (DW). LMA was calculated as the ratio of the DW to the leaf disk area (g m^{-2}).

Photosynthetic activity and light energy dissipation mechanisms were assessed through measurements of leaf pigments, including chlorophylls, flavonoids, and anthocyanins, as well as chlorophyll fluorescence parameters. Measurements were taken on 3–6 different plants for each species on sunny days between 10:00 a.m. and 12:00 a.m. across five seasonal periods: November 2022, January 2023, May 2023, August 2023, and September 2023. The air temperature (Tair) and relative humidity (RH) during the measurement campaigns were obtained by the schoolchildren under the supervision of their teachers using the school meteorological station, while the photosynthetic photon flux density (PPFD) was recorded through the fluorometer sensor (Table 2).

Table 2. Mean values (± standard error) of photosynthetic photon flux density (PPFD), air temperature (Tair), and relative humidity (RH) recorded every half hour during the chlorophyll fluorescence measurements (~10–12 a.m.).

	Sampling Date	PPFD (μmol m^{-2} s^{-1})	Tair (°C)	RH (%)
1	23 November 2022	1530 ± 6	14.98 ± 0.38	68.60 ± 0.21
2	25 January 2023	1249 ± 22	9.22 ± 0.11	53.83 ± 0.83
3	8 May 2023	1537 ± 37	20.25 ± 0.16	63.50 ± 0.65
4	21 August 2023	1782 ± 15	35.95 ± 0.96	35.75 ± 2.46
5	27 September 2023	1791 ± 9	26.98 ± 0.38	46.50 ± 3.18

Leaf pigment indices were determined on fully expanded and well-exposed leaves of various Mediterranean shrubland species using a portable MPM-100 multi-pigment meter (Opti-Science Inc., Hudson, NH, USA). Measurements of leaf fluorescence (F) ratios were conducted to evaluate indices of anthocyanin (Anth, F660nm/F525nm) and flavonoid (Flav, F660nm/F325nm) content, as well as leaf transmission in the far red and near infrared to assess the chlorophyll content index (Chl, T850nm/T720mm). On the same leaves used for pigment determination, chlorophyll fluorescence was measured utilizing a pulse amplitude modulated fluorometer (Mini-PAM; Heinz Walz GmbH, Effeltrich, Germany). The leaves were pre-darkened for approximately 30 min, and the maximum quantum efficiency of PSII photochemistry (Fv/Fm) was evaluated using the formula $(Fm - F_0)/Fm$, where the difference between the maximum (Fm) and minimum (F_0) fluorescence yield emitted by the leaves in the dark-adapted state represents the variable fluorescence (Fv). The PPFD of the saturating light flash used to determine Fm was approximately 8000 μmol m^{-2} s^{-1}. Measurements of the steady-state effective quantum yield of photosystem II (PSII) photochemistry in the light Y(II) and both the light-regulated Y(NPQ) and non-regulated Y(NO) non-photochemical energy dissipation at PSII were conducted on fully expanded and well-exposed leaves under ambient PPFD, RH, and Tair values (Table 2).

The Y(II) value was calculated using the formula (Fm′ − F′)/Fm′, where Fm′ represents the maximum fluorescence yield with all PSII reaction centres in the reduced state, obtained by superimposing a saturating light flash during exposure to actinic light, and F′ is the fluorescence at the current state of PSII reaction centres during actinic illumination. The Fv′/Fm′ value was determined as (Fm′ − F_0′)/Fm′, where F_0′ is the minimum fluorescence yield in the light-adapted state, measured immediately after switching off the actinic light and applying a weak far-red light to preferentially excite PSI and force electrons to drain from PSII. The Y(NPQ) and Y(NO) values were calculated as (F/Fm′)/(F/Fm) and F/Fm, respectively. The Y(II) values were utilised to calculate the rate of linear electron transfer (ETR) [34], multiplying Y(II) by the incident PPFD and then correcting for the actual fraction of absorbed light and the relative distribution of absorbed light between the two photosystems, assumed to be 0.5 and 0.84, respectively [35]. The chlorophyll fluorescence decrease ratio (RFd) was calculated as (Fm − Fs)/Fs; this parameter is often used as an indicator of the "vitality" of plants and evolution of stress [36].

The schoolchildren were instructed on how to conduct the aforementioned ecophysiological measurements and understand their meaning, while being supervised by an expert throughout the activity. Additionally, the fluorometer was equipped with a screen that enabled the students to verify when the fluorescence signal was stable over time, indicating when a measurement could be taken. Any measurements that were unstable were repeated until valid values were obtained. Finally, a post-processing of the data was performed to discard any values collected under unstable conditions or in inappropriate light exposure compared to external conditions. The schoolchildren were educated about this validation process and the importance of handling scientific data to ensure its reliability. This enabled us to compare the measurements taken throughout the monitoring period and correlate them with the meteorological trends.

2.6. Statistical Analysis

Statistical analyses were performed using the STATISTICA software package 8.0 (StatSoft for Windows, 1998, Tulsa, OK, USA). Data on leaf functional traits were analysed independently using a one-way ANOVA for the factors of species and time. Pairwise multiple comparisons within each factor were conducted using the Fisher LSD method. Linear regressions were employed to test the relationships between individual variables of interest, with determination coefficients (R^2) and significance levels (*p*-value) reported.

3. Results

An OAL has been established in the urban park of Villa Corridi, involving 19 schoolchildren aged 10 years and 22 schoolchildren aged 11–12 years, along with their teachers, in two separate events. A dedicated project on the iNaturalist website was created, featuring georeferenced photos of selected evergreen shrub and tree species taken by the teachers (https://www.inaturalist.org/projects/la-biodiversita-del-parco-villa-corridi, accessed on 18 October 2024; Figure S5). Additionally, an educational classroom herbarium has been developed and is now available for teaching activities (Figures S6 and S7). The meteorological data recorded by the schoolchildren and their teachers using the school's meteorological station are presented in Figure 1b. These data reveal the typical trends of the Mediterranean climate, characterised by significant seasonal fluctuations in both temperature and precipitation. Average daily temperatures fell below 10 °C almost only during winter, while precipitation was mostly concentrated in autumn, with few occurrences in other seasons. The meteorological trends were used to assess the interaction between climate and ecophysiological leaf traits (i.e., chlorophyll fluorescence parameters and leaf pigment indices) of the studied species. Seasonal variations in light energy dissipation within the Mediterranean shrubland formation, through both photochemical and non-photochemical processes, are illustrated in Figure 2. The average values for the 11 plant species in this formation revealed a significant decrease in Y(II) during January and August, which was associated with an increase in Y(NO) and Y(NPQ), respectively. This seasonal change in light energy dissipation processes was reflected in the ETR values, which exhibited the lowest seasonal values in January and August, corresponding to the minimum and maximum seasonal daily air temperatures, respectively (Figure 1b). The reduced ETR in January was linked to the lowest seasonal values of both Fv/Fm and Rfd, while the pigment indices (i.e., Chl, Flv, and Anth) reached their maximum seasonal values (Table 3). In contrast, the decreased ETR values in August were associated with only a slight decline in Fv/Fm and a significant reduction in the Chl index (Table 3). Significant interspecific differences in energy dissipation processes (Figure 3a–c), Fv/Fm (Figure 4a), ETR (Figure 4b), and Chl index (Figure 4c) were observed among the 11 Mediterranean species. The significance levels (*p*-values) for the factors of time and species are reported in Tables S1 and S2. All the species tended to increase Y(NPQ) and Y(NO) in summer and winter, respectively (Figure 3b,c). The Fv/Fm values showed only a slight decrease in winter and summer for most species (Figure 4a). This was reflected in a transient decrease in Y(II) (Figure 3a) and ETR (Figure 4b) in all the species, except for *Q. ilex* and *P. angustifolia* in August, and of *Q. ilex*, *P. angustifolia* and *C. creticus* in January. A partial or total recovery of ETR after the summer inhibition was observed in September for all the species, except for *L. lucidum* (Figure 4b). The species that exhibited a reduced ETR in August also showed a decrease in the Chl index (Figure 4c). Consequently, a positive relationship between ETR and Chl index was observed in August (R = 0.83; $p < 0.01$), with *Q. ilex, P. angustifolia, P. halepensis,* and *P. lentiscus* displaying higher values of both ETR and Chl compared to the other species (Figure 5a). These sclerophyllous species were also characterised by higher LMA values, leading to a positive relationship between ETR and LMA (R = 0.89; $p < 0.01$) (Figure 5b). Conversely, in January, a positive relationship between ETR and Anth index was observed (R = 0.89; $p < 0.05$), with *P. angustifolia, Q. ilex, P. halepensis,* and *C. creticus* showing the highest Anth index values (Figure 5c).

Table 3. Seasonal variation of electron transport rate (ETR), maximum photochemical efficiency of PSII (Fv/Fm), chlorophyll fluorescence decrease ratio (Rfd), chlorophyll index (Chl), flavonoid index (Flav), and anthocyanin index (Anth). Values are the average of the 11 plant species of the Mediterranean shrub formation of Villa Corridi urban park. Data were analysed independently by one-way ANOVA. For each parameter, data followed by different letters in the same line are significantly different for the significance level indicated.

	Nov	Jan	May	Aug	Sep	*p*-Value
ETR (μmol m^{-2} s^{-1})	95.3 a	55.2 b	102.5 a	65.6 b	100.2 a	<0.0001
Fv/Fm	0.816 a	0.740 d	0.792 b	0.773 c	0.795 b	<0.0001
Rfd	2.61 c	1.50 d	3.57 b	3.85 b	4.62 a	<0.0001
Chl	1.21 a	1.25 a	1.10 b	0.91 c	0.96 c	<0.0001
Flav	1.07 c	1.33 a	1.20 b	0.98 c	1.19 b	<0.0001
Anth	0.02 c	0.16 a	0.11 b	0.01 c	0.00 c	<0.0001

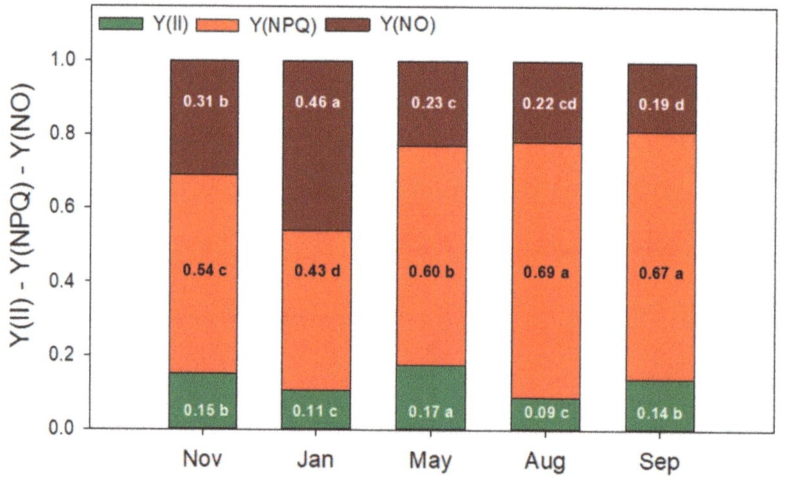

Figure 2. Seasonal variation of steady-state effective quantum yield of photosystem II (PSII) photochemistry in the light Y(II), light-regulated Y(NPQ), and non-regulated Y(NO) non-photochemical energy dissipation at PSII. Values are the average of the 11 plant species of the Mediterranean shrub formation of Villa Corridi urban park. Data were analysed independently by one-way ANOVA. For each parameter, data followed by different letters in the same line are significantly different ($p \leq 0.05$, Fisher's LSD).

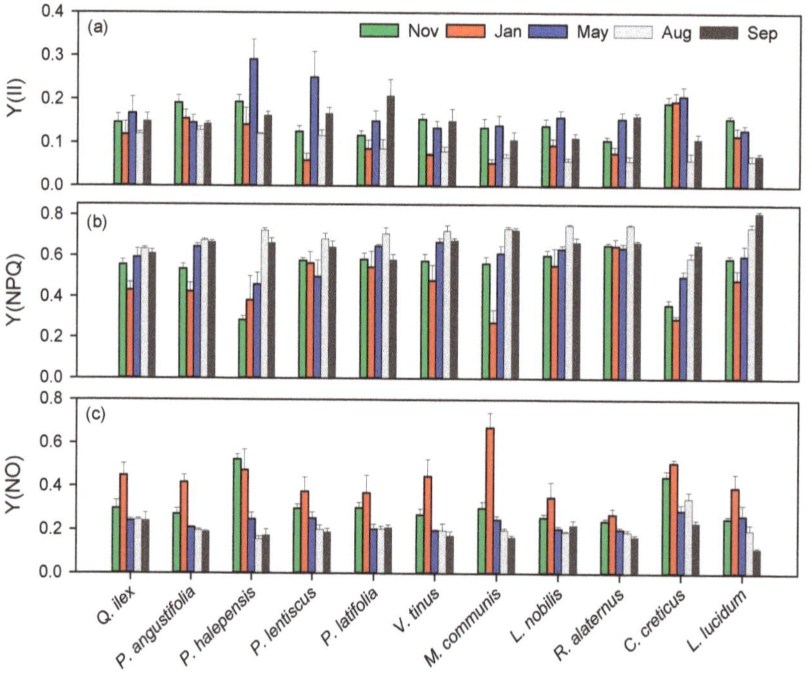

Figure 3. Seasonal variation of (**a**) steady-state effective quantum yield of photosystem II (PSII) photochemistry in the light Y(II), (**b**) light-regulated Y(NPQ), and (**c**) non-regulated Y(NO) non-photochemical energy dissipation at PSII in the 11 plant species of the Mediterranean shrub formation of the park of Villa Corridi. Data were analysed independently by one-way ANOVA for the variable factors of species and time (Table S1).

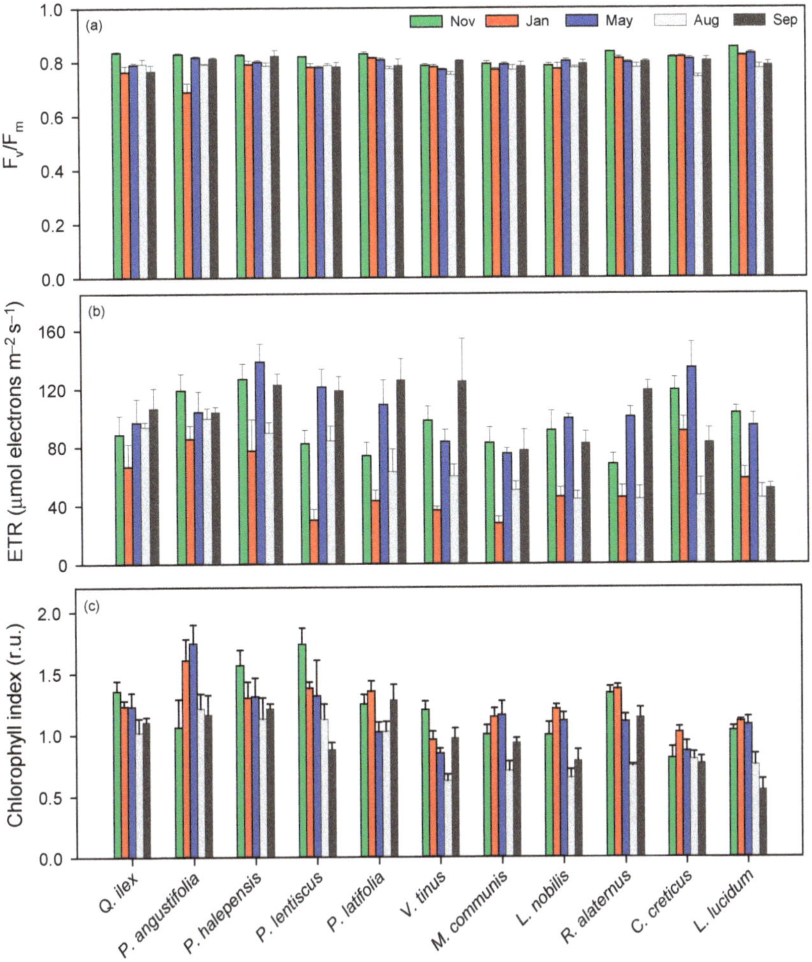

Figure 4. Seasonal variation of (**a**) maximum photochemical efficiency of photosystem II in the dark (Fv/Fm), (**b**) electron transport rate (ETR), and (**c**) chlorophyll index (Chl) in the 11 plant species of the Mediterranean shrub formation of Villa Corridi park. Data were analysed independently by one-way ANOVA for the variable factors of species and time (Table S2).

Approximately 200 schoolchildren (ages 9 to 11 years) and 16 teachers participated in additional training sessions in the OAL, identifying and collecting samples of wild woody species, measuring their photosynthetic performance, and gathering meteorological data.

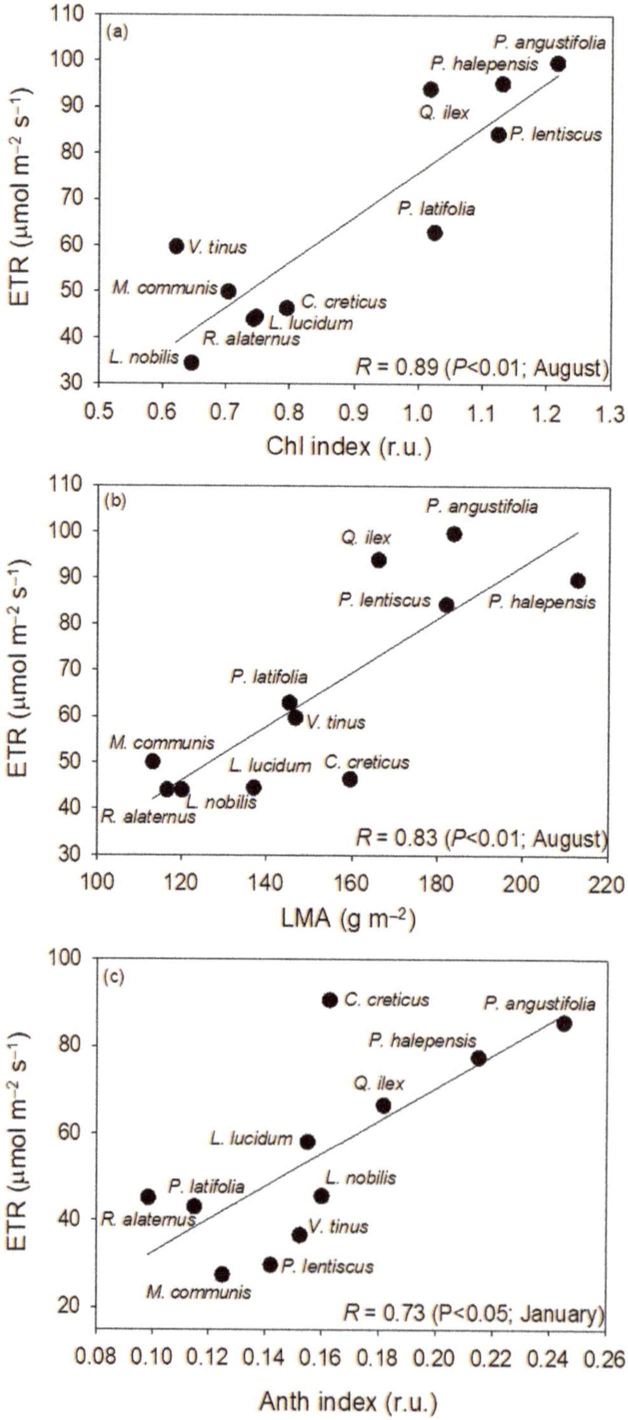

Figure 5. Linear relationships between (**a**) electron transport rate (ETR) and chlorophyll index (Chl) in August; (**b**) ETR and leaf mass per area (LMA) in August; and (**c**) ETR and anthocyanin index (Anth) in January in the 11 plant species of the Mediterranean shrub formation of the park of Villa Corridi.

4. Discussion

The significance of involving children in citizen science activities related to the environment has been extensively discussed and promoted [17]. Active participation in ecological research is particularly valuable for enhancing children's understanding of this scientific field [37]. These activities can also serve as a source of reliable data, provided that quality is ensured through meticulous verification [38,39]. Schoolchildren participating in the experimental campaigns in the park of Villa Corridi searched for selected species and recorded the morphophysiological leaf traits. This location has proven to be an effective living lab for studying wild plant biodiversity and its ecophysiological responses to seasonal environmental changes. This approach, combined with classroom lessons, enabled the participants to (1) enhance their understanding of their surrounding environment and foster a sense of belonging; (2) acquire foundational knowledge of botanical terminology used to identify eight common Mediterranean species and key ecological processes; (3) analyse the connections between biotic and abiotic components of the ecosystem; (4) improve skills such as communication, classification, measurement, inference, and prediction; (5) increase their awareness of the key role of urban green infrastructure, particularly Mediterranean evergreen species, in promoting human well-being and supporting ecosystems; and (6) practice group-collaboration while completing a task. The involvement of schoolchildren and their teachers in citizen science activities related to wild plant species and the environment is particularly significant. During the activities, the students demonstrated their ability to connect a picture and a simple list of diagnostic characters to a living plant. This can be viewed as a preliminary acquisition of basic plant identification knowledge, including habitus (tree vs. small or large shrub), phyllotaxy (opposite vs. alternate), leaf shape (simple vs. compound), leaf colour (concolorous vs. discolorous), leaf margins (entire vs. dentate), leaf scent (scented vs. grassy), and leaf indumentum (hairy vs. glabrous). Previous research has indicated that the general public may not fully understand what biodiversity is and how it relates to ecology [40]. Moreover, children may primarily be attracted to cultivated plants, while wild and less showy species may only capture the interest of researchers [41]. It appears that some children experience a form of "plant blindness" toward native species [14]. The fruitful engagement of children in these research activities through gamification and their direct involvement in specimen collection and analysis has demonstrated that this approach can effectively foster a connection between young students and science. The creation of a photographic atlas on the iNaturalist app, featuring georeferenced observations of the studied species, can also be useful for classroom activities [42]. Each project on iNaturalist includes a map displaying the georeferenced observations from the explored area, allowing for a review of observed biodiversity and familiarization with a technology that simplifies and makes the process of learning the names and identification of wild organisms enjoyable. Additionally, creating a classroom herbarium is an excellent method for teaching botany in schools [43]. Classroom discussions encouraged students to connect the observed plant species with recorded meteorological data and measurements related to plant ecophysiology obtained during activities in the park. This approach, which deviates from the typical classroom routine, exemplifies a multidisciplinary science teaching module [44]. Finally, it is important to emphasise that the informal green space within the park of Villa Corridi has facilitated the establishment of the OAL due to the abundant presence of native wild species. The significance of informal green spaces at the urban level, in terms of the ecosystem services they provide, has been extensively documented [12], particularly in comparison to formal green areas [45].

The direct engagement of schoolchildren with scientific instruments commonly used by researchers to measure photosynthesis and other leaf traits represents a novel approach to educational activities. Under expert supervision, they utilised the chlorophyll fluorescence technique as a powerful tool to assess the photosynthetic performance and health status of Mediterranean shrubland species [46]. This experience enabled the participants to (1) apply theoretical knowledge from school modules related to photosynthesis through hands-on field experience; (2) understand the instrumentation and techniques that scientists employ to generate replicable and reliable scientific data; and (3) evaluate how the photosynthetic performance of individual Mediterranean species is influenced by the environmental conditions in which they thrive (e.g., solar radiation, precipitation, air temperature, and relative humidity), as recorded by the school's meteorological station. The results indicated a reduction in the PSII photochemical efficiency and electron transport rate of the Mediterranean shrub vegetation in both August and January. This finding highlights a diminished capacity to dissipate light energy through photochemistry during the hot, dry summer and the cold winter periods. In August, daily air temperatures reached their maximum seasonal values, coinciding with a lack of significant precipitation until the end of the month. Conversely, in January, minimum temperatures dropped to near 0 °C. The decreased Y(II) was linked to an increase in the quantum yield of light-dependent (August) and light-independent (January) non-photochemical fluorescence quenching. These processes represent the energy dissipated as heat through regulated and non-regulated energy dissipation pathways, respectively [47]. These results clearly indicate that the fate of absorbed radiation energy at the PSII level varies between winter and summer, alternating the direction of excessive absorbed energy between regulated and non-regulated energy dissipation pathways. Specifically, during the hot and dry summer months, the Mediterranean maquis formation effectively protected PSII by dissipating excessive light energy through regulated non-photochemical processes. In contrast, during winter, the plants increased the proportion of energy lost through harmful non-regulated dissipative pathways [48]. It has been hypothesised that high Y(NO) values are associated with a longer lifetime of energy excitation, which can lead to the formation of reactive oxygen species [49]. This could lead to photoinhibition and photodamage, as indicated by the reduced Fv/Fm and vitality index observed in winter. Furthermore, the increase in the Flav and Anth indices in January suggests a crucial role for antioxidant defences in mitigating oxidative stress and reducing the risk of photoinhibition [50,51]. Conversely, during the hot and dry summer period, the reduction in photochemical efficiency was associated with an increase in Y(NPQ), highlighting the ability of Mediterranean shrub species to safely dissipate excessive absorbed energy as heat without exhibiting chronic photoinhibition signals. Nevertheless, the temporary decrease in Y(II) and ETR during summer confirms that high air temperatures and excessive solar radiation can lead to a reduction in photosynthetic capacity in Mediterranean vegetation [52,53]. Seasonal variations in photosynthetic performance and energy dissipation processes were examined in relation to changes in leaf pigment composition. The reduced ETR in August was linked to a decrease in the Chl index, indicating a rearrangement of the photosynthetic apparatus in response to multi-stress conditions typical of summer. Previous studies have shown that a reduction in chlorophyll content, which is associated with increased leaf reflectance, may help Mediterranean vegetation mitigate excessive light energy by reducing intercepted solar radiation [54,55]. These results highlight the capacity of evergreen Mediterranean maquis to safely dissipate excess energy as heat (i.e., thermal energy dissipation) during the hot and dry summer months. This mechanism helps prevent chronic photodamage to the photosynthetic apparatus, enabling the recovery of photosynthetic performance in September when environmental conditions become milder. Conversely, cold winter periods can negatively impact Mediterranean vegetation by inducing chronic photoinhibition processes and activating the antioxidant system, as indicated by the increase in the

Anth index. Research has shown that Mediterranean species experience two main stress periods: winter and summer [56]. Low winter temperatures result in chronic photoinhibition, while the hot and dry summer months lead to dynamic photoinhibition and a reduction in pigment content. This suggests that summer photoprotection may arise from a combination of increased non-radiative energy dissipation, alternative electron sinks, and a higher carotenoid-to-chlorophyll ratio [56].

The other question addressed with the schoolchildren was whether there were interspecific differences in how plants respond to environmental changes. In fact, species-specific adaptive responses to seasonal environmental changes were observed among the studied Mediterranean species, depending on their peculiar morphophysiological characteristics. Specifically, sclerophyllous species exhibiting the highest LMA values, such as *Q. ilex*, *P. angustifolia*, *P. lentiscus*, and *P. pinea*, were able to maintain a higher ETR and Chl index during the summer compared to the other species investigated. This suggests a strong tolerance of their photosynthetic apparatus to photoinhibition under hot and dry environmental conditions [57–59]. Notably, the Mediterranean sclerophyllous species can achieve a high capacity for CO_2 uptake despite their large LMA, resulting in high productivity [23]. Moreover, some of these species, such as *P. angustifolia*, are considered well suited for the multi-stress conditions typical of urban environments [60] and have been proposed as promising ornamental species for gardening and landscaping in Mediterranean areas [61]. An exception was represented by *L. lucidum*, which, despite exhibiting high LMA values, showed a limited ability to withstand summer stress conditions and did not recover its photochemical efficiency in September. This observation may be attributed to the optimal conditions for this species, which are found in the central and southern regions of China, where a humid subtropical climate prevails [62]. *Quercus Ilex*, *P. angustifolia*, and *P. pinea* maintained higher ETR values during the winter months compared to the other studied species, highlighting their capacity to mitigate cold stress. This ability is also associated with the synthesis of anthocyanins, which function as an antioxidant system to prevent chronic photoinhibition [50,51]. These findings confirm the significant growth plasticity and adaptability of these species to both summer drought and winter frost events in Mediterranean regions [63]. Conversely, the decline in ETR observed in *P. lentiscus* during the winter suggests a sensitivity of this species to cold stress [64]. A distinctive seasonal behaviour was observed in *C. creticus*, a semi-deciduous species of the Mediterranean maquis. *Cistus* species can shed a substantial portion of their leaves during the summer, showing a combination of drought-tolerance and drought-avoidance strategies [65]. According to our data, these species generally exhibited a downregulation of PSII associated with a reduction in leaf chlorophyll concentrations and a rearrangement in the structure of the light harvesting complex to minimise potential photoinhibition during the summer [66]. This was followed by a recovery of physiological performance during the autumn–winter period [67], thereby confirming the resilience of this semi-deciduous species to the Mediterranean climate [68].

Overall, these data support the plasticity and adaptability of shrubland species to both summer and winter stress periods, making them excellent candidates for NbS in urban areas located in coastal Mediterranean regions.

5. Conclusions

This study represents a successful pilot citizen science initiative aimed at engaging young students in biodiversity monitoring and raising their awareness of environmental issues. The results demonstrated the feasibility of an Open-Air Laboratory, where students explored their everyday environment from a new perspective and observed aspects they had previously overlooked, effectively gamifying science. This approach possibly not only heightened their awareness of the importance of biodiversity but also fostered a greater sense of belonging. Furthermore, the direct involvement of schoolchildren in collecting ecophysiological data on the mechanisms of plant resilience to environmental constraints was successfully achieved. Ultimately, the methodology employed in this study can serve as a pilot model for similar future initiatives.

Supplementary Materials: The following supporting information can be downloaded at https://www.mdpi.com/article/10.3390/land14020423/s1: Figure S1: Climate chart for the city of Livorno; Figures S2 and S3: Cards with species descriptions; Figure S4: Schoolchildren involvement in project activities; Figure S5: The main page of the project created in the iNaturalist website; Figures S6 and S7: Classroom herbarium specimens; and Tables S1 and S2: Statistical analyses.

Author Contributions: Conceptualisation, A.S. and V.L.; methodology, A.S. and V.L.; validation, A.S., V.L. and F.B.; formal analysis, A.S.; investigation, A.S. and V.L.; resources, A.S., V.L., G.R., L.P. and L.M.; data curation, A.S., V.L., G.R. and L.P.; writing—original draft preparation, A.S. and V.L.; writing—review and editing, G.R., L.P., A.M., L.M., R.P., E.T. and F.B.; supervision, A.S.; project administration, A.S., V.L., G.R., L.P., L.M., R.P., E.T. and A.M.; and funding acquisition, A.S. and F.B. All authors have read and agreed to the published version of the manuscript.

Funding: Project funded under the National Recovery and Resilience Plan (NRRP), Mission 4 Component 2 Investment 1.4—Call for tender No. 3138 of 16 December 2021, rectified by Decree n. 3175 of 18 December 2021 of Italian Ministry of University and Research funded by the European Union—NextGenerationEU; Award Number: Project code CN_00000033, Concession Decree No. 1034 of 17 June 2022 adopted by the Italian Ministry of University and Research, CUP B83C22002930006, Project title "National Biodiversity Future Center—NBFC".

Data Availability Statement: The data presented in this study are available on request from the corresponding author. The data are not publicly available due to privacy restrictions.

Acknowledgments: The authors are grateful to Maria Salvatrice Oriti (Istituto Comprensivo G. Bartolena) for having permitted and encouraged the activities with schoolchildren. We also thank Giuseppe Mario Lanini and Giacomo Tagliaferri (CNR IBE) for the educational activities with the meteorological station and Diana Prince (WWF Livorno) for supporting the activities in the park. Finally, we thank the Municipality of Livorno for the management of the park. This work is in memory of Battista Scartazza who instilled in his sons a deep love for nature.

Conflicts of Interest: The authors declare no conflicts of interest.

References

1. Weber, C. Ecosystem services provided by urban vegetation: A literature review. In Proceedings of the 11th Urban Environment Symposium (UES), Karlsruhe, Germany, 16–19 September 2012; Rauch, S., Morrison, G., Norra, S., Schleicher, L., Eds.; Springer: Dordrecht, The Netherlands, 2012; pp. 119–131.
2. Pereira, P.; Yin, C.; Hua, T. Nature-based solutions, ecosystem services, disservices, and impacts on well-being in urban environments. *Curr. Opin. Environ. Sci. Health* **2023**, *33*, 100465. [CrossRef]
3. Langemeyer, J.; Gómez-Baggethun, E. Urban biodiversity and ecosystem services. In *Urban Biodiversity: From Research to Practice*, 1st ed.; Niemelä, J., Ossola, A., Eds.; Routledge: New York, NY, USA, 2018; pp. 36–53.
4. Pinho, P.; Moretti, M.; Luz, A.C.; Grilo, F.; Vieira, J.; Luís, L.; Rosalino, L.J.; Martins-Loução, M.A.; Santos-Reis, M.; Correia, O.; et al. Biodiversity as support for ecosystem services and human wellbeing. In *The Urban Frest: Cultivating Green Infrastructure for People and the Environment*; Pearlmutter, D., Calfapietra, C., Samson, R., O'Brien, L., Krajter Ostoić, S., Sanesi, G., Alonso del Amo, R., Eds.; Springer: Cham, Switzerland, 2017; pp. 67–78. [CrossRef]
5. Grote, R.; Samson, R.; Alonso, R.; Amorim, J.H.; Cariñanos, P.; Churkina, G.; Fares, S.; Le Thiec, D.; Niinemets, Ü.; Mikkelsen, T.N.; et al. Functional traits of urban trees: Air pollution mitigation potential. *Front. Ecol. Environ.* **2016**, *14*, 543–550. [CrossRef]
6. Livesley, S.J.; McPherson, G.M.; Calfapietra, C. The urban forest and ecosystem services: Impacts on urban water, heat, and pollution cycles at the tree, street, and city scale. *J. Environ. Qual.* **2016**, *45*, 119–124. [CrossRef] [PubMed]
7. World Health Organization. *Urban Green Spaces and Health*; WHO Regional Office for Europe: Copenhagen, Denmark, 2016.
8. Dávila, N.; Moura, E.; Versieux, L.M.; Carvalho, F.A.; Calvente, A. Urban forest fragments as a living laboratory for teaching botany: An example from federal University of Rio Grande do Norte. *Brazil. Sys. Bot.* **2021**, *46*, 6–17. [CrossRef]

9. Gallotti, G.; Santo, M.A.; Apostolidou, I.; Alessandri, J.; Armigliato, A.; Basu, B.; Debele, S.; Domeneghetti, A.; Gonzalez-Ollauri, A.; Kumar, P.; et al. On the management of nature-based solutions in open-air laboratories: New insights and future perspectives. *Resources* **2021**, *10*, 36. [CrossRef]
10. Stanford, H.R.; Hurley, J.; Garrard, G.E.; Kirk, H. Exploring the secret gardens of the city: An assessment of human-nature interactions on informal green space using citizen science data. *Urban For. Urban Green.* **2024**, *98*, 128414. [CrossRef]
11. Vereecken, N.J.; Weekers, T.; Marshall, L.; D'Haeseleer, J.; Cuypers, M.; Pauly, A.; Pasau, B.; Leclercq, N.; Tshibungu, A.; Molenberg, J.; et al. Five years of citizen science and standardised field surveys in an informal urban green space reveal a threatened Eden for wild bees in Brussels, Belgium. *Insect Conserv. Divers.* **2021**, *14*, 868–876. [CrossRef]
12. Luo, S.; Patuano, A. Multiple ecosystem services of informal green spaces: A literature review. *Urban For. Urban Green.* **2023**, *81*, 127849. [CrossRef]
13. Beery, T.; Jørgensen, K.A. Children in nature: Sensory engagement and the experience of biodiversity. *Environ. Educ. Res.* **2018**, *24*, 13–25. [CrossRef]
14. Ballouard, J.M.; Brischoux, F.; Bonnet, X. Children prioritize virtual exotic biodiversity over local biodiversity. *PLoS ONE* **2011**, *6*, e23152. [CrossRef] [PubMed]
15. Robinson, L.D.; Cawthray, J.L.; West, S.E.; Bonn, A.; Ansine, J. Ten principles of citizen science. In *Citizen Science. Innovation in Open Science, Society and Policy*; Hecker, S., Haklay, M., Bowser, A., Makuch, Z., Vogel, J., Bonn, A., Eds.; UCL Press: London, UK, 2018; pp. 27–40.
16. Pocock, M.J.O.; Roy, H.E.; August, T.; Kuria, A.; Barasa, F.; Bett, J.; Githiru, M.; Kairo, J.; Kimani, J.; Kinuthia, W.; et al. Developing the global potential of citizen science: Assessing opportunities that benefit people, society and the environment in East Africa. *J. Appl. Ecol.* **2019**, *56*, 274–281. [CrossRef]
17. Makuch, K.; Aczel, M. Children and citizen science. In *Citizen Science: Innovation in Open Science, Society and Policy*; Hecker, S., Haklay, M., Bowser, A., Makuch, Z., Vogel, J., Bonn, A., Eds.; UCL Press: London, UK, 2018; pp. 391–409.
18. MacPhail, V.J.; Colla, S.R. Power of the People: A Review of Citizen Science Programs for Conservation. *Biol. Conserv.* **2020**, *249*, 108739. [CrossRef]
19. Fritz, S.; See, L.; Carlson, T.; Haklay, M.; Oliver, J.L.; Fraisl, D.; Mondardini, R.; Brocklehurst, M.; Shanley, L.A.; Schade, S.; et al. Citizen science and the United Nations Sustainable Development Goals. *Nat. Sustain.* **2019**, *2*, 922–930. [CrossRef]
20. Wünsche, H.; Schimmler, S. Citizen science and the dissolution of inequalities in scientific knowledge production. In *Challenges of Digital Inequality—Digital Education, Digital Work, Digital Life, Proceedings of the Weizenbaum Conference 2019, Berlin, Germany, 16–17 May 2019*; Weizenbaum Institute: Berlin, Germany, 2019. [CrossRef]
21. Dorst, H.; Van Der Jagt, A.; Raven, R.; Runhaar, H. Urban greening through nature-based solutions—Key characteristics of an emerging concept. *Sustain. Cities Soc.* **2019**, *49*, 101620. [CrossRef]
22. Galmés, J.; Medrano, H.; Flexas, J. Photosynthetic limitations in response to water stress and recovery in Mediterranean plants with different growth forms. *New Phytol.* **2007**, *17*, 81–93. [CrossRef]
23. Flexas, J.; Díaz-Espejo, A.; Gago, J.; Gallé, A.; Galmés, J.; Gulías, J.; Medrano, H. Photosynthetic limitations in Mediterranean plants: A review. *Environ. Exp. Bot.* **2014**, *103*, 12–23. [CrossRef]
24. Di Pirro, E.; Sallustio, L.; Castellar, J.A.; Sgrigna, G.; Marchetti, M.; Lasserre, B. Facing multiple environmental challenges through maximizing the co-benefits of nature-based solutions at a national scale in Italy. *Forests* **2022**, *13*, 548. [CrossRef]
25. Schwarz, N.; Moretti, M.; Bugalho, M.N.; Davies, Z.G.; Haase, D.; Hack, J.; Hof, A.; Melero, Y.; Pett, T.J.; Knapp, S. Understanding biodiversity-ecosystem service relationships in urban areas: A comprehensive literature review. *Ecosyst. Serv.* **2017**, *27*, 161–171. [CrossRef]
26. Kamarudin, M.Z.; Mat Noor, M.S.A. To what extent do primary school children understand photosynthesis?: A study of children's drawings with follow-up interviews. *Education 3-13* **2024**, 1–16. [CrossRef]
27. Gaiotto, A.; Tonon, S.; Santovito, G. The scientific method in the teaching of life sciences in primary school, The plants and their seasonal changes. In Proceedings of the EDULEARN13 Proceedings, Barcelona, Spain, 1–3 July 2013; pp. 4226–4235.
28. Miczajka, V.L.; Klein, A.-M.; Pufal, G. Elementary school children contribute to environmental research as citizen scientists. *PLoS ONE* **2015**, *10*, e0143229. [CrossRef] [PubMed]
29. Kottek, M.; Grieser, J.; Beck, C.; Rudolf, B.; Rubel, F. World map of the Köppen-Geiger climate classification updated. *Meteorol. Zeitschrift* **2006**, *15*, 259–263. [CrossRef]
30. Zepner, L.; Karrasch, P.; Wiemann, F.; Bernard, L. ClimateCharts.net—An interactive climate analysis web platform. *Int. J. Digit. Earth* **2020**, *14*, 338–356. [CrossRef]
31. Rivas-Martínez, S.; Rivas-Sáenz, S.; Penas-Merino, A. Worldwide bioclimatic classification system. *Glob. Geobot.* **2011**, *1*, 1–634.
32. Williamson, K.B.; Kang, Y.P.; Steele, J.L.; Gunderman, R.B. The art of asking: Teaching through questioning. *Acad. Radiol.* **2002**, *9*, 1419–1422. [CrossRef] [PubMed]

33. Bowser, A.; Hansen, D.; He, Y.; Boston, C.; Reid, M.; Gunnell, L.; Preece, J. Using gamification to inspire new citizen science volunteers. In Proceedings of the First International Conference on Gameful Design, Research, and Applications; Association for Computing Machinery, New York, NY, USA, 2–4 October 2013; pp. 18–25. [CrossRef]
34. Genty, B.; Briantais, J.; Baker, N.R. The relationship between the quantum yield of photosynthetic electron transport and quenching of chlorophyll fluorescence. *Biochim. Biophys. Acta* **1989**, *990*, 87–92. [CrossRef]
35. Maxwell, K.; Johnson, G.N. Chlorophyll fluorescence—A practical guide. *J. Exp. Bot.* **2000**, *51*, 659–668. [CrossRef] [PubMed]
36. Roháček, K.; Barták, M. Technique of the modulated chlorophyll fluorescence: Basic concepts, useful parameters, and some applications. *Photosynthetica* **1999**, *37*, 339–363. [CrossRef]
37. Michael, J. Where's the evidence that active learning works? *Adv. Physiol. Educ.* **2006**, *30*, 159–167. [CrossRef]
38. Shirk, J.L.; Ballard, H.L.; Wilderman, C.C.; Phillips, T.; Wiggins, A.; Jordan, R.; McCallie, E.; Minarchek, M.; Lewenstein, B.V.; Krasny, M.E.; et al. Public participation in scientific research: A framework for deliberate design. *Ecol. Soc.* **2012**, *17*, 29. [CrossRef]
39. Castagneyrol, B.; Valdés-Correcher, E.; Bourdin, A.; Barbaro, L.; Bouriaud, O.; Branco, M.; Centenaro, G.; Csóka, G.; Duduman, M.; Dulaurent, A.; et al. Can school children support ecological research? Lessons from the Oak bodyguard citizen science project. *Citizen Sci. Theory Pract.* **2019**, *5*, 1–11. [CrossRef]
40. Lindemann-Matthies, P.; Bose, E. How many species are there? Public understanding and awareness of biodiversity in Switzerland. *Hum. Ecol.* **2008**, *36*, 731–742. [CrossRef]
41. Muratet, A.; Pellegrini, P.; Dufour, A.-B.; Arrif, T.; Chiron, F. Perception and knowledge of plant diversity among urban park users. *Landsc. Urban Plan.* **2015**, *137*, 95–106. [CrossRef]
42. Chozas, S.; Nunes, A.; Serrano, H.C.; Ascensão, F.; Tapia, S.; Máguas, C.; Branquinho, C. Rescuing botany: Using citizen-science and mobile apps in the classroom and beyond. *Npj Biodivers.* **2023**, *2*, 6. [CrossRef]
43. Flannery, M.C. The road to herbaria: Teaching and learning about biology, aesthetics, and the history of botany. *J. Biosci.* **2023**, *48*, 58. [CrossRef]
44. Doyle, M.; Bozzone, D. Multidisciplinary teaching providing undergraduates with the skills to integrate knowledge and tackle 'big' questions. *J. Center Interdiscip. Teach. Learn.* **2018**, *7*, 12–17.
45. Pietrzyk-Kaszyńska, A.; Czepkiewicz, M.; Kronenberg, J. Eliciting non-monetary values of formal and informal urban green spaces using public participation GIS. *Landsc. Urban Plan.* **2017**, *160*, 85–95. [CrossRef]
46. Murchie, E.H.; Lawson, T. Chlorophyll fluorescence analysis: A guide to good practice and understanding some new applications. *J. Exp. Bot.* **2013**, *64*, 3983–3998. [CrossRef] [PubMed]
47. Kramer, D.M.; Johnson, G.; Kiirats, O.; Edwards, G.E. New fluorescence parameters for the determination of QA redox state and excitation energy fluxes. *Photosyn. Res.* **2004**, *79*, 209–218. [CrossRef] [PubMed]
48. Shtai, W.; Asensio, D.; Kadison, A.E.; Schwarz, M.; Raifer, B.; Andreotti, C.; Hammerle, A.; Zanotelli, D.; Haas, F.; Niedrist, G.; et al. Soil water availability modulates the response of grapevine leaf gas exchange and PSII traits to a simulated heat wave. *Plant Soil* **2024**, *501*, 537–554. [CrossRef]
49. Samson, G.; Bonin, L.; Maire, V. Dynamics of regulated YNPQ and non-regulated YNO energy dissipation in sunflower leaves exposed to sinusoidal lights. *Photosyn. Res.* **2019**, *141*, 315–330. [CrossRef]
50. Hughes, N.M.; Neufeld, H.S.; Burkey, K.O. Functional role of anthocyanins in high-light winter leaves of the evergreen herb *Galax urceolata*. *New Phytol.* **2005**, *168*, 575–587. [CrossRef]
51. Landi, M.; Tattini, M.; Gould, K.S. Multiple functional roles of anthocyanins in plant-environment interactions. *Environ. Exp. Bot.* **2015**, *119*, 4–17. [CrossRef]
52. Chaves, M.M.; Pereira, J.S.; Maroco, J.; Rodrigues, M.L.; Ricardo, C.P.P.; Osório, M.L.; Carvalho, I.; Faria, T.; Pinheiro, C. How plants cope with water stress in the field? Photosynthesis and growth. *Ann. Bot.* **2002**, *89*, 907–916. [CrossRef] [PubMed]
53. Peñuelas, J.; Munné-Bosch, S.; Llusià, J.; Filella, I. Leaf reflectance and photo-and antioxidant protection in field-grown summer-stressed Phillyrea angustifolia. Optical signals of oxidative stress? *New Phyt.* **2004**, *162*, 115–124. [CrossRef]
54. Munné-Bosch, S.; Alegre, L. Changes in carotenoids, tocopherols and diterpenes during drought and recovery, and the biological significance of chlorophyll loss in *Rosmarinus officinalis* plants. *Planta* **2000**, *210*, 925–931. [CrossRef]
55. Gratani, L.; Varone, L. Adaptive photosynthetic strategies of the Mediterranean maquis species according to their origin. *Photosynthetica* **2004**, *42*, 551–558. [CrossRef]
56. Baquedano, F.J.; Castillo, F.J. Drought tolerance in the Mediterranean species *Quercus coccifera*, *Quercus ilex*, *Pinus halepensis*, and *Juniperus phoenicea*. *Photosynthetica* **2007**, *45*, 229–238. [CrossRef]
57. Martín-Sánchez, R.; Peguero-Pina, J.J.; Alonso-Forn, D.; Ferrio, J.P.; Sancho-Knapik, D.; Gil-Pelegrín, E. Summer and winter can equally stress holm oak (*Quercus ilex* L.) in Mediterranean areas: A physiological view. *Flora* **2022**, *290*, 152058. [CrossRef]
58. Larcher, W. Temperature stress and survival ability of Mediterranean sclerophyllous plants. *Plant Biosyst.* **2000**, *134*, 279–295. [CrossRef]
59. Iozia, L.M.; Crisafulli, V.; Varone, L. Climatic variations along an aridity gradient drive significant trait intraspecific variability in Mediterranean plant species. *J. Arid Environ.* **2023**, *217*, 105042. [CrossRef]

60. Pellegrini, E.; Cotrozzi, L.; Neri, L.; Baraldi, R.; Carrari, E.; Nali, C.; Lorenzini, G.; Paoletti, E.; Hoshika, Y. Stress markers and physiochemical responses of the Mediterranean shrub *Phillyrea angustifolia* under current and future drought and ozone scenarios. *Environ. Res.* **2021**, *201*, 111615. [CrossRef] [PubMed]
61. Álvarez, S.; Gómez-Bellot, M.J.; Acosta-Motos, J.R.; Sánchez-Blanco, M.J. Application of deficit irrigation in *Phillyrea angustifolia* for landscaping purposes. *Agric. Water Manag.* **2019**, *218*, 193–202. [CrossRef]
62. Montti, L.; Píriz Carrillo, V.; Gutiérrez-Angonose, J.; Gasparri, I.; Aragón, R.; Grau, R. The role of bioclimatic features, landscape configuration and historical land use in the invasion of an Asian tree in subtropical Argentina. *Landsc. Ecol.* **2017**, *32*, 2167–2185. [CrossRef]
63. Di Filippo, A.; Baliva, M.; Brunetti, M.; Di Fiore, L. Long-term tree-ring response to drought and frost in two *Pinus halepensis* populations growing under contrasting environmental conditions in Peninsular Italy. *Forests* **2021**, *12*, 305. [CrossRef]
64. Stefi, A.L.; Papaioannou, V.; Nikou, T.; Halabalaki, M.; Vassilacopoulou, D.; Christodoulakis, N.S. Heat and cold-stressed individuals of *Pistacia lentiscus* (Mastic Tree) do modify their secreting profile. *Plants* **2022**, *11*, 3290. [CrossRef] [PubMed]
65. Harley, P.C.; Tenhunen, J.D.; Beyschlag, W.; Lange, O.L. Seasonal changes in net photosynthesis rates and photosynthetic capacity in leaves of *Cistus salvifolius*, a European Mediterranean semi-deciduous shrub. *Oecologia* **1987**, *74*, 380–388. [CrossRef]
66. Grant, O.M.; Tronina, Ł.; García-Plazaola, J.I.; Esteban, R.; Pereira, J.S.; Chaves, M.M. Resilience of a semi-deciduous shrub, *Cistus salvifolius*, to severe summer drought and heat stress. *Funct. Plant Biol.* **2014**, *42*, 219–228. [CrossRef]
67. Catoni, R.; Gratani, L.; Varone, L. Physiological, morphological and anatomical trait variations between winter and summer leaves of *Cistus* species. *Flora* **2012**, *207*, 442–449. [CrossRef]
68. Puglielli, G.; Catoni, R.; Spoletini, A.; Varone, L.; Gratani, L. Short-term physiological plasticity: Trade-off between drought and recovery responses in three Mediterranean *Cistus* species. *Ecol. Evol.* **2017**, *7*, 10880–10889. [CrossRef]

land

MDPI

Article

Exploring the Possibilities of Implementing the ALS-Based 3-30-300 Concept for Urban Green Space Management in Small Municipalities

Krzysztof Mitelsztedt , Mariusz Ciesielski *, Tomasz Hycza , Marek Lisańczuk , Kacper Guderski , Sylwia Kurpiewska and Krzysztof Korzeniewski

Department of Geomatics, Forest Research Institute, Sękocin Stary, Braci Leśnej 3, 05-090 Raszyn, Poland; k.mitelsztedt@ibles.waw.pl (K.M.); t.hycza@ibles.waw.pl (T.H.); m.lisanczuk@ibles.waw.pl (M.L.); k.guderski@ibles.waw.pl (K.G.); s.kurpiewska@ibles.waw.pl (S.K.); k.korzeniewski@ibles.waw.pl (K.K.)

* Correspondence: m.ciesielski@ibles.waw.pl

Abstract: This study examines the applicability of the 3-30-300 rule in five medium-sized Polish municipalities. The rule states that residents should be able to see at least three trees from their homes, neighborhoods should have at least 30% tree canopy coverage, and public green spaces should be within 300 m. The method proposed in this study shows that the tree visibility component of the 3-30-300 concept is the most fluctuating index, and it strongly depends on the settings of the algorithm parameter, as well as on the placement of artificially generated observers. This demonstrates the complexity of the issue and the need to further specify the nuances of the 3-30-300 rule. The work shows that all variables of the 3-30-300 rule can be calculated based on publicly available data, such as point clouds, which are increasingly being made available for free for research and implementation purposes. The study concludes that the proposed solution is effective in assessing the availability of green spaces and emphasizes the need for qualitative improvements in the management of urban green spaces. While the 3-30-300 rule can serve as the foundation for future urban planning, complementary strategies are needed to ensure long-term sustainability and better access to green spaces.

Keywords: 3-30-300; urban greenery; green space; tree cover; tree visibility; airborne laser scanning

1. Introduction

Numerous actions are being undertaken to adapt cities to climate change with the aim of improving the living conditions of their inhabitants. Many of them have also been implemented as part of European Union law [1]. The World Health Organization (WHO) has developed several indicators regarding this topic, e.g., health indicators related to air quality and extreme weather events, indicators of access to green spaces, and indicators of health adaptation [1]. One of the tools for adapting cities to the effects of climate change and improving the living conditions of a growing number of inhabitants is the appropriate design of urban green spaces—from local to city-wide spatial levels [2].

Greenery in urban areas is an inseparable element of the urban fabric and composition [3]. The concept of urban greenery, in the broadest sense, encompasses all biologically active areas in a city. It includes both managed green spaces (parks, squares, green areas, etc.) and unmanaged greenery (forests, isolated green spaces, woodlots, etc.) [4]. In a narrower sense, urban greenery can only be understood as the population of trees in the above-mentioned areas. Regardless of the definition, the role of urban greenery in shaping urban space and improving the quality of life of residents is significant [5].

According to the concept of ecosystem services, urban greenery can provide services in four categories: provisioning, regulating, cultural, and supporting [6,7]. In urban areas, timber-harvesting activities related to local timber resources are of marginal interest [8]. As surveys show, city residents value the regulating and cultural services provided by green areas and trees in cities much more [9]. Urban greenery reduces the risk of the urban heat island effect by providing shading, evapotranspiration, and a reduction in thermal amplitude [10,11]. Trees also reduce surface runoff and, thus, the occurrence of local flooding due to heavy rainfall [12]. In addition, urban greenery, especially trees, effectively reduces air pollution, including particulate matter [13]. Reducing pollution decreases mortality due to respiratory diseases [14]. Another regulatory service provided by greenery is noise reduction [15]. It has been known for many years that urban greenery has a positive effect on the health and well-being of city residents [16]. Numerous studies at the intersection of various disciplines and scientific fields have shown that contact with nature and with trees improves one's psychophysical state [17], reduces stress [18], and reduces the risk of depression [19]. This positive effect has been demonstrated in recent years during the COVID-19 pandemic [20]. In addition to direct contact with nature, the mere possibility of seeing green spaces, e.g., through a window, also has a positive effect on the lives of residents [21]. Urban greenery also provides important cultural services related to aesthetic values, complements the urban composition, and inspires and forms the identity of the place [22]. The cultural service of recreation in green spaces is extremely valuable for city dwellers [23,24], especially as access to green spaces in highly urbanized areas varies greatly, and there are inequalities in access for different social groups [25]. This is due to investment pressure, the need for new residential areas, low-quality spatial planning, and a lack of urban standards and land values. These factors often cause green areas to change their purpose to other types of land use. Green areas are becoming a deficit asset, the demand for which significantly exceeds supply.

All of the ecosystem services mentioned above directly or indirectly affect the well-being of citizens in place of their residence and the standard of living of society [26]. It should be mentioned that urban greenery is an important element of a nature-based solution (NbS), as it can provide a wide range of ecosystem services if properly placed and planned. Therefore, the appropriate identification of green resources and making decisions that support the modification of resources to better adapt to climate change should be one of the challenges for city managers. NbS-based activities provide benefits for residents and nature, and one of the elements that NbSs could improve is fair and adequate access to urban green spaces for city residents [27,28]. This statement was reflected in the 2030 Agenda—the world development strategy adopted by the United Nations in 2015 [29]. One of the goals was defined as ensuring easy and universal access to safe green areas, taking into account various groups that use the space, including women, children, the elderly, and the disabled. The Polish Ministry of Climate and Environment also emphasized the availability of green spaces as a criterion that can be used by local authorities for various strategic decisions in the handbook published in January 2022, which contains natural and climatic indicators for sustainable development [30].

Existing studies on the thresholds defining the sufficient amount of greenery per capita, the availability of managed green areas, and their size differ both in the literature and in strategic and planning documents [31]. The guidelines of the World Health Organization (WHO) indicate that areas located at a distance of more than 300 m from forests and managed green areas should be treated as "excluded" [32,33]. Furthermore, this distance is counted only from forests and green areas whose area is larger than 1 ha. This results from numerous studies on the minimum size of a green area that has appropriate characteristics to provide certain ecosystem services [34,35]. Kabisch et al. [36] determined the demand and supply of urban green spaces by calculating the number of people within a radius of 300 and 500 m around green areas of over 2 ha. Cardinali et al. [37] indicated the distance of up to 100 m as one with a positive impact on health and physical activity. Greenery in the vicinity of 500–1100 m from the place of residence also has a positive effect on health. Different guidelines were included by Polish legislation on urban standards when amending the Act on Spatial Planning and Development [38]. According to the provisions, the distance from public green areas should be less than 1500 m in the case of public greenery with a total area of no less than 3 ha, and it should be within 3000 m for areas over 20 ha. In addition to quantitative issues, researchers point to the need to create high-quality green areas [39].

Sustainable urban management that involves actions related to climate change adaptations can make use of synthetic indices. In particular, one such indicator, the "3-30-300 rule" presented by Konijnendijk et al. [40], has gained prominence in recent years. This guideline aims to ensure equitable access to green spaces by establishing the following thresholds: (1) at least three well-established trees within sight of every home, school, and workplace, (2) at least 30% tree cover in every neighborhood, and (3) no more than a 300 m distance from any residence to the nearest public green space. What is notable is that the author of the concept did not specify a calculation method for the above components, which means that different research teams are currently proposing their own solutions. To assess access to green spaces using the "3-30-300" rule, some authors [41] have conducted questionnaire surveys. Others have conducted visual and spatial analyses of images provided by respondents [42], online databases and services [43], and remote sensing data [44].

The individual components of the 3-30-300 index have also been realized differently. Nieuwenhuijsen et al. [45] assessed the greenness of the window view with a question about whether participants could see trees from their homes. They combined this information with data on the number of trees within 15 m of a residence using a central geodatabase. The locations of buildings from OpenStreetMap and the locations of trees from urban databases were used to determine the visibility of trees in 20 m [46] or 30 m buffers [47]. Ling [43] determined the total number of trees of a "decent" size that could be seen from the front window of each residential building using Google Street View, Google satellite images, or 360-degree views provided by an estate agent. Zhang et al. [42] calculated window views using the Green View Index of photos taken by respondents from their most viewed windows with the best viewing angle. Daland [44] proposed viewshed tool analysis, an algorithm that calculates all areas visible from a given point based on a raster elevation model.

The second component of the 3-30-300 rule, i.e., canopy cover, was also determined in various manners. Nieuwenhuijsen et al. [45] used the Normalized Difference Vegetation Index (NDVI) in 500 m buffers to estimate canopy cover in the direct vicinity. Zhang et al. [42] applied Sentinel-2 data with a spatial resolution of 10 m. Ling [43] used a similar method that was implemented in i-Tree Canopy. This tool randomly places points on Google Earth images, allowing the user to decide on their class. Browning et al. [48] recommended Digital Surface Models (DSMs) from LiDAR (Light Detection and Ranging) in conjunction with high-resolution imagery and high-accuracy land cover maps for the measurement of green space locations and the determination of canopy cover. Daland [44] applied zonal statistics to calculate the area occupied by trees based on the canopy height model.

The proximity of green areas has been a subject of consideration for many years. Nieuwenhuijsen et al. [45] calculated the linear distance to the nearest green area. Ling [43] determined the walking distance from the place of residence to the nearest boundary or path of a park or green space using Google Maps. The park or green space had to be public, free of charge, and at least 1 hectare in size. Daland [44] performed a cost distance analysis by calculating the shortest weighted distance from each cell of a cost surface raster to the nearest source location. Browning et al. [48] proposed several available types of Geographic Information System (GIS) data as a source of information to identify the locations of urban green spaces in order to determine their spatial relationships. Distances can be operationalized based on "as the crow flies" (i.e., Euclidean) or with more sophisticated analysis, which takes into account transportation networks (e.g., streets and junctions) and access points (e.g., centroids or entry points for the urban green space). Another approach was used by Zhang et al. [42], who calculated the average NDVI value within a 300 m buffer around check-in locations using the Google Earth Engine platform.

Taking into account the range of various approaches to realizing the 3-30-300 concept, it seems that remote sensing (RS) can be an appropriate tool for the determination of urban-greenery-related indices at the level of both individual trees and their features, as well as groups of trees [49]. Segmentation of individual trees was performed by Fekete and Cserep [50], who successfully determined single trees' species, location, health status, crown parameters, and other features that were relevant from the point of view of managing greenery resources. Tree detection could also be performed for smaller areas using terrestrial laser scanning data [51], while mobile laser scanning could be used for linear objects [52]. Entire cities are mapped using airborne laser scanning (ALS) [53] or aerial image content analysis [54]. Depending on the complexity of the data and methods, the tree detection accuracy ranged from 62 to 100% [55,56].

Until now, the above-mentioned projects have mainly been undertaken for larger cities [46,57]. Nevertheless, the share of small and medium-sized municipalities in the structure of urban areas in the European Union (EU) is considerable. In 2023, there were almost 8000 towns in the EU, home to 95 million people. Two-thirds of these were densely populated, i.e., with 1500 inhabitants/km^2 and with populations between 5000 and 50,000 inhabitants. The population of towns was distributed almost evenly across the three size categories; just over a third lived in small towns (5000–10,000 inhabitants) and medium-sized towns (10,000–25,000), while just under a third lived in bigger towns (>25,000) [58]. We can assume that large urban agglomerations have both policies and resources for urban greenery management. However, this cannot always be the case in every town, the main reason for this being the limited budget and human resources available. On that account, there the reason for investigating the possibilities of transferring the 3-30-300 concept to the needs of smaller residential areas emerges.

In this paper, we present an original method for identifying urban green space resources in cities of different sizes based on the 3-30-300 concept. Our approach utilized free and open data such as OpenStreetMap and the increasingly collected public and available non-commercial point clouds [59]. The major focus of our work was to test a new approach to determining the '3' component of tree visibility from buildings. We proposed a method based on so-called observers, i.e., artificially generated positions of tree observations. Since we also have information on the number of observers, we were tempted to identify so-called 'high-priority trees' that could be of greater value to a larger number of residents.

The proposed approach offers the opportunity to replicate the methodology in analyses for cities that do not have the financial and human resources to collect LiDAR data. In contrast to most work on this topic, we mainly use only one type of remote sensing data as a source. The presented solution makes it possible to identify places where there is a deficit of green spaces, and, therefore, appropriate management measures are needed. Moreover, the objective nature of the proposed methods allows comparisons between cities/regions thanks to the application of ALS data.

2. Materials and Methods

2.1. General Concept

The main idea was to develop a solution that would make it possible to efficiently determine the availability of urban green spaces according to the guidelines of the 3-30-300 concept. The solution was intended to be universally applicable but dedicated foremost to smaller municipalities, where detailed inventory data on urban vegetation are scarce. Our proposal is based on the use of ALS point clouds and the OpenStreetMap database (Figure 1). The application of GIS and RS tools and data should provide standardized and objective results so that municipalities can be compared according to the guidelines of the 3-30-300 principle. In the following subsections, the implementation of the individual components of the 3-30-300 concept is presented in detail.

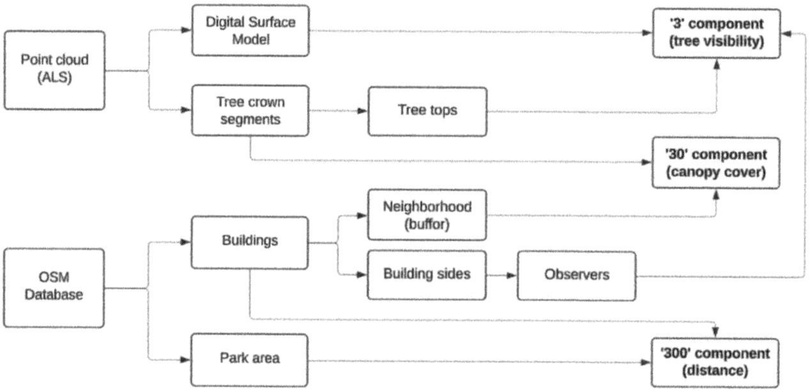

Figure 1. General concept for implementing the 3-30-300 method.

2.2. Study Area

The analysis focused on five towns: Świdnik, Wyszków, Czempiń, Jasień, and Mrocza (Figure 2, Table 1) which are representative of the demographic and urban structure in Poland. Almost 38 million people lived in Poland at the end of 2023. Around 60% of them were in the country's 1013 towns. The urban structure is dominated by towns with less than 5000 inhabitants (44%), while this accounts for only 3% of people living in towns. The total number of small towns (5000–10,000 inhabitants) and medium-sized towns (10,000–25,000) is similar—around 400—but almost half of all urban inhabitants of Polish towns are agglomerated in these groups. Every third inhabitant lives in a city with a population of more than 50,000 [60].

Table 1. Selected characteristics of the towns presented in this study.

Characteristics	Czempiń	Jasień	Mrocza	Świdnik	Wyszków
Area [km^2]	5.0	3.6	4.8	20.4	20.9
Population	4137	5127	4044	36,806	26,042
Population density [people/km^2]	824	1420	846	1803	1252
Green areas [%]	0.8	0.6	1.07	8.1	2.1
Forest cover [%]	1.6	2.2	6.1	9.6	3.4

The selection of these cities resulted from the fact that approximately 80% of Polish cities have fewer than 50,000 inhabitants. As mentioned before, similar towns usually do not have sufficient tools for greenery monitoring, which makes them proper study objects. The general characteristics of the study areas are presented in Table 1 and Figure 2.

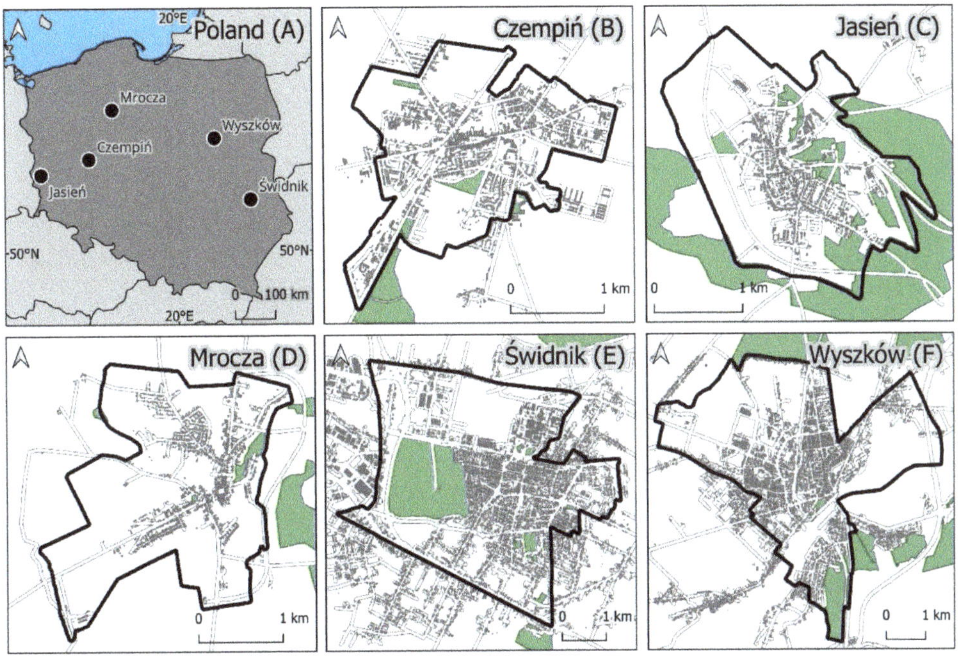

Figure 2. Location of selected study areas (**B–F**) in Poland (**A**). Green spaces over one hectare analyzed in this work are marked in green.

2.3. Data Source

Lidar data from the Head Office of Geodesy and Cartography [61] were primarily used for the analyses in this project. Airborne scanning missions were carried out in different time periods: in Czempiń and Wyszków in 2011, in Świdnik in 2011–2013, in Mrocza in 2015, and in Jasień in 2021. The point cloud density of about 4 points/m^2 was consistent across all of the towns. The points' average elevation error was 0.15 m. The point clouds had already been classified by the data provider according to the LAS 1.2 standard [62]. In this study, ALS data were used for treetop detection, single-crown segmentation, and DSM generation, which served as a basis for analyzing the components of the 3-30-300 principle. Other remote sensing data were derived from the OpenStreetMap database [63], namely, the spatial layers of residential buildings and green areas. The green areas were additionally verified by city officials.

2.4. Point Cloud Processing

The point clouds were processed in the *R* environment version 4.2.3 [64] using the *lidR* library [65,66]. In order to detect the treetops, the height of the point clouds was normalized in the first step. This process removed the influence of the terrain on the measurements above the ground. This made it possible to extract and compare the heights of the aboveground vegetation. For this purpose, a k-nearest neighbor (KNN) approach with an inverse-distance weighting (IDW) algorithm for spatial interpolation was used.

The point clouds were further processed with a local maximum filter. This was a *lidR* algorithm for individual tree detection inspired by the work of Popescu and Wynne [67]. The threshold for trees was set to a height of at least 5 m. The resulting layer of points representing treetops was then manually evaluated. To improve the accuracy, the positions of treetops along the roads were manually adjusted.

In addition, a rasterized Digital Surface Model (DSM) with a resolution of 0.5 m was created from the point cloud. This layer was created based on the Pit-free algorithm developed by Khosravipour et al. [68], which is based on the calculation of a series of classical triangulations at different heights.

2.4.1. Tree Visibility

In this study, the term "visibility" should be understood as a direct linear and unobstructed line of sight between the observer and the treetop. The potential visibility of the trees was simulated from each side of each building. The visibility of a tree was determined in two phases.

In phase "A", superfluous vertices, i.e., those whose removal did not change the overall orthogonal shape of the building, were deleted from the polygons. The polygons were then converted into lines representing the contours of the building. These lines were then cut into individual sections, and their lengths were calculated. Eventually, only lines whose length was greater than half the average length of all sections of a particular building were included in further analysis. In this way, only the main walls forming the generic body of the building were identified. This operation was carried out to reduce the noise caused by tiny sections of a building facade where the presence of a potential window seemed unlikely.

In the next step, an extruded point was created at the center of each section at a distance of 0.5 m perpendicular to the elevation line and at a height of 1.7 m, which corresponds to the height of an average person (white dots in Figure 3A). Henceforth, throughout the article, such points are referred to as "observers". An additional advantage of this approach is that it is easy to identify so-called "blind" observers (red dots in Figure 3A), i.e., places where a building has no windows because it shares a wall with another building.

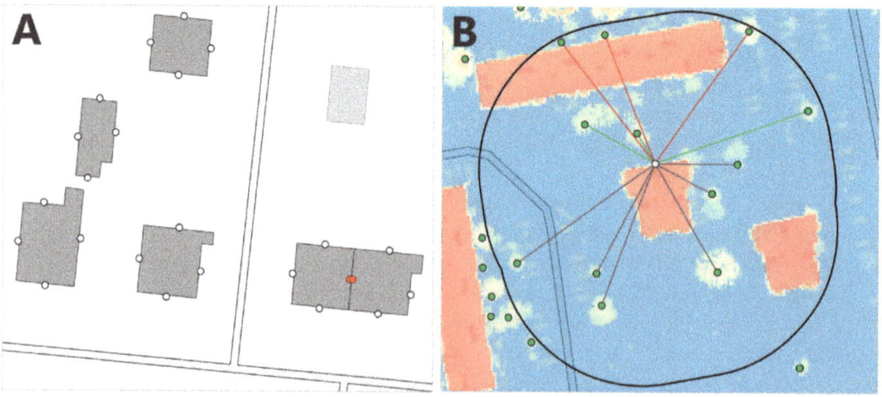

Figure 3. The visibility of a tree was determined in two phases. In phase "(**A**)", the observers (white dots) and "blind" observers (red dots) were determined. In phase "(**B**)", the lines of sight were delineated in the field of view (black outline). A tree (green dot) could be visible (green line), obscured by other objects (red line), or not visible from this side of the building (gray line).

Phase "B" (Figure 3B) consisted of drawing a line of sight between the treetops and each observer. If the line crossed the building (gray lines in Figure 3B), this meant that the observer could not see the tree. For the line where the view was not obstructed by the building, a terrain profile was extracted from the DSM raster. At this stage, the visibility condition was met if the treetop was higher than other objects in the line of sight.

An observer passed the visibility constraint if the condition was met for three or more trees. Per-building results were distinguished depending on the assumption of how many observers in a building needed to see three trees in order for the entire building to be considered as fulfilling the "3" component. Due to the lack of a precise definition of this component, we proposed three variants: (I) at least 1 observer sees at least three trees, (II) at least half of the observers (each) see at least three trees, and (III) more than 90% of the observers fulfill the criterion of the first component.

Additionally, a backward analysis was conducted to identify high-priority trees. By reversing the visibility condition, we recorded how many observers could see each tree from how many buildings. With this slight alteration, we were able to identify the priority trees to which the community should pay special attention.

2.4.2. Canopy Coverage

The *sf* library for the R programming language was used to determine the "30" component [69,70]. The tree canopy cover was generated from the point cloud by transforming points of class 5 according to the LAS specification [62] into a raster with a 0.5 m resolution. Next, the proportion of pixels representing tree crowns in the total area of every building's neighborhood was calculated. As it is subjective to clearly define the appropriate neighborhood radius, variants with five distances were chosen: 30, 60, 90, 120, and 150 m from the residential buildings. The final result was the percentage of the area occupied by tree canopies in these buffer zones.

2.4.3. Distance from a Public Green Space

As in the previous analysis, R version 4.2.3 [64] was used to determine the "300" component. First, the building and green space layers were downloaded from the OpenStreetMap database using the *osmdata* library [71]. Then, the polygons were processed using the *sf* library [69,70]. Only green areas larger than one hectare were considered. The choice of green spaces with a larger area was related to the guidelines developed by the WHO [32]. The datasets of potential green space polygons were then cross-checked with city councils to ensure that they were publicly accessible to all residents. The green spaces were classified into three categories: small (1–5 ha), medium (5–10 ha), and large (over 10 ha). Then, the Euclidean distance was measured from the outline of every building to the boundaries of the green space with a straight planar line. This gave us a slightly better insight into the distance that residents have to travel to the different types of green spaces, although the interpretation remained relatively simple.

3. Results

The results themselves were not the primary aim of this study, but rather the presentation of the possibilities that arise when RS data are included in the analytical process. The visualization and final analysis of the results were conducted using the QGIS 3.34 software [72]. The possibilities of data interpretation are presented together with the results in the following subsections on the individual components of the 3-30-30 concept. In addition, a synthetic measure was proposed in different variants, as the lack of the rigid constraints of the 3-30-300 method makes the results susceptible to the threshold criteria.

3.1. The "3" Component—Tree Visibility

Table 2 shows the share of buildings from which at least three trees are visible. In all towns, more than 80% of the buildings met this criterion in variant I. The higher the percentage of observers inside every building who were able to see at least three trees, the lower the percentage of buildings that fulfilled this criterion. In the strictest variant, where it was assumed that >90% of observers see the corresponding number of trees, the percentage of buildings that met this criterion fell to around 25%.

Table 2. Percentage of buildings meeting the criterion of visibility of at least three trees depending on the proportion of observers.

Town	Tree Visibility Component for the Buildings * [%]		
	I	II	III
Czempiń	88.0	55.4	26.8
Jasień	88.6	53.9	26.2
Mrocza	89.7	55.4	28.2
Świdnik	80.7	49.9	23.6
Wyszków	80.5	51.1	25.8

* Variants: (I) at least one observer can see three or more trees; (II) at least half of the observers (each) can see three or more trees; (III) at least 90% of the observers (each) see at least three trees.

Interesting results were provided by our modification of the visibility component that determined the high-priority trees. Recording information about how many observers can see a particular tree provides an interesting tool for urban decision makers. In this way, we can point out trees and areas (Figure 4) that need extra attention due to the large number of people who see these trees from their windows every day. For example, the proportion of trees seen by 10 or more observers from their windows was less than 1% in four out of five towns. The highest value was in Czempiń, where it reached 4.5%. The maximum values (i.e., the total number of observers) of high-priority trees ranged from 15 in Jasień to 26 in Świdnik.

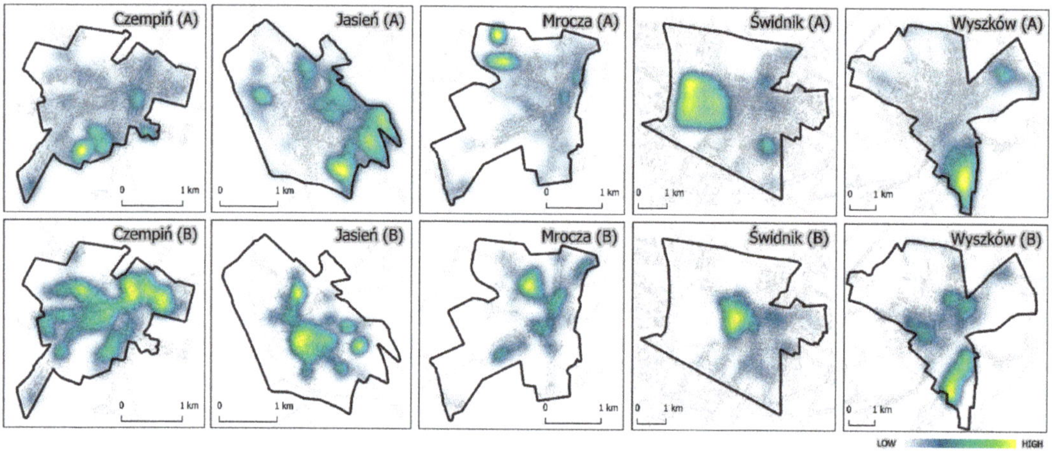

Figure 4. An example of the effect that arises when the number of trees (**A**) is weighted by the number of observers who can see the trees (**B**).

3.2. The "30" Component—Tree Canopy Cover

The tree canopy cover in the vicinity of the building was analyzed in five different distance zones (Table 3). Only in Jasień did the enlargement of the buffer around the building lead to a significant increase in the percentage of buildings meeting this criterion. The greatest gain (6.7 percentage points) was observed when transitioning from 30 to 60 m. In other towns, the change in the size of the neighborhood area had hardly any effect on the criterion. In Wyszków, the proportion of buildings that met the criterion was highest (7.9%) when the neighborhood had a 30 m buffer zone. In the other neighborhood variants, the percentage of buildings that met the 30% canopy cover criterion was highest in Jasień.

In Mrocza, Świdnik, and Czempiń, the percentage of buildings in the different variants did not exceed 0.4, 1.4, and 2.9%, respectively.

Table 3. Percentage of buildings that meet the tree canopy cover criterion with the different neighborhood size options.

Town	Share of Buildings Meeting the Criterion of Tree Canopy Cover [%]				
	30 m	60 m	90 m	120 m	150 m
Czempiń	2.5	2.4	2.7	2.7	2.9
Jasień	4.0	10.7	12.6	13.8	16.0
Mrocza	0.1	0.1	0.4	0.1	0.0
Świdnik	1.1	1.1	1.2	1.3	1.4
Wyszków	7.9	8.6	8.4	8.4	8.5

3.3. The "300" Component—Access to Green Areas

The analysis showed that 21.8% of the residential buildings in Wyszków and even 77.3% of the buildings in Jasień were located within a radius of 300 m from green areas of more than one hectare (Table 4). If we took into account the distance from larger green areas, the number of buildings that fulfilled this criterion decreased significantly. When considering green spaces of more than 10 hectares, only Świdnik fulfilled this criteria. A peculiar situation concerned Mrocza, where, despite having two large green areas, there were no residential buildings within 300 m.

Table 4. Percentage of buildings meeting the distance criterion in different variants.

Town	Share of Buildings Meeting the Criterion of Distance [%]					
	Size of Green Area			Total Number of Green Areas		
	≥1 ha	≥5 ha	≥10 ha	≥1 ha	≥5 ha	≥10 ha
Czempiń	38.3	29.6	5.0	4	1	3
Jasień	77.3	53.6	14.9	7	3	2
Mrocza	46.7	32.6	0.0	5	2	2
Świdnik	61.9	32.3	31.2	18	2	5
Wyszków	21.8	12.9	9.5	10	9	1

3.4. Results of the 3-30-300 Rule

The two variants of the results for the analyzed towns shown in Table 5 differ in the criterion for the number of observers and the area of green spaces. The neighborhood criterion for the calculation of the tree canopy cover was left unchanged, as this variable did not depend on the distance assumptions for the selected towns. Regardless of the variant, the percentage of buildings meeting all 3-30-300 criteria was not high in the towns analyzed. In variant I, 14.2% of the buildings in Jasień met all three components, while in Mrocza, no residential buildings met all criteria. The vast majority of buildings (around 80–90%) met one or two criteria. The highest percentage of buildings that did not fulfill any of the criteria was recorded in Wyszków (15.8%).

The tightening of the criteria in variant II and the consideration of a higher number of observers in the building (at least 50%) who had to see at least three trees, as well as the distance from green areas of at least 10 ha, lead to a decrease in the percentage of buildings that fulfilled the criteria. In Jasień, the percentage of buildings meeting the three criteria fell by almost 10 percentage points, while in Wyszków, it only decreased by around 0.6 percentage points. To better illustrate the results, the distribution of the individual components of the 3-30-300 rule in the two variants discussed is shown in Figure 5.

Table 5. Percentage of residential buildings in relation to the number of criteria met.

Town	Variant I *				Variant II **			
	0	1	2	3	0	1	2	3
Czempiń	7.6	57.7	32.6	2.1	42.24	52.57	4.84	0.35
Jasień	2.7	26.9	56.2	14.2	36.00	47.84	11.52	4.64
Mrocza	4.8	54.0	41.2	0.0	44.60	55.40	0.00	0.00
Świdnik	8.5	40.2	50.0	1.3	35.51	47.53	16.03	0.93
Wyszków	15.8	67.4	16.7	4.9	45.44	44.32	5.98	4.27

* At least one observer in the building sees three trees, the canopy cover is indicated at a distance of 150 m from the building, and the distance to green areas is at least 1 ha; ** at least 50% of the observers in the building see three trees, the canopy cover is indicated at a distance of 150 m from the building, and the distance to green areas is at least 10 ha.

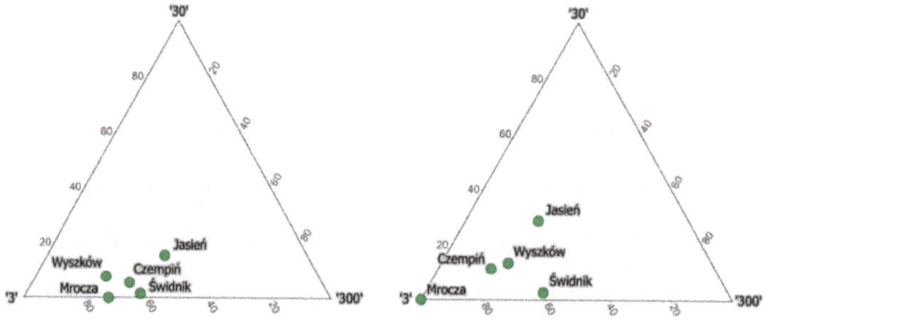

Figure 5. The performance of the towns in each component in variant I (**left**) and variant II (**right**).

The criterion evaluation is also shown in aggregated form on the grid level (Figures 6 and 7). A square in the grid provides information on the number of criteria for the majority of buildings that fall within that square. In all towns, 6.1% of the squares in variant I were counted as meeting all criteria, 36.7% were counted as fulfilling two criteria, 51.9% were counted as fulfilling one criterion, and 5.7% were counted as not fulfilling any criteria. In variant II, the corresponding percentages were 3.4%, 10.4%, 52.8%, and 33.5%.

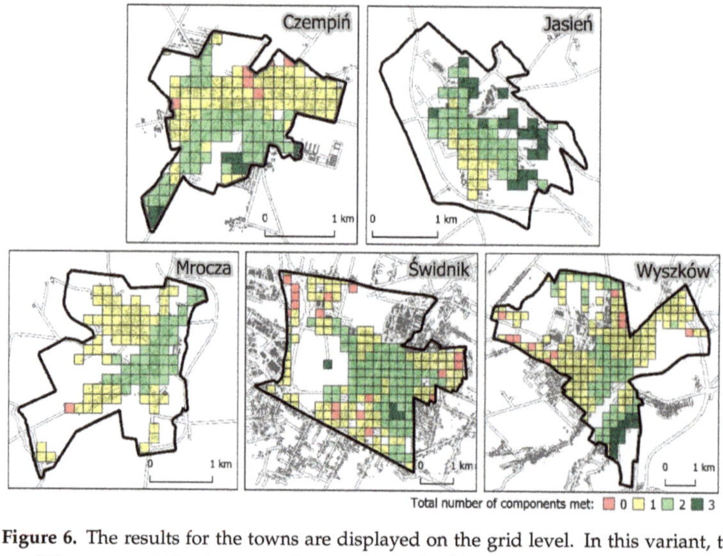

Figure 6. The results for the towns are displayed on the grid level. In this variant, the visibility condition was met if at least one observer in the building could see three or more trees and the building was within 300 m of a green area of 1 ha or larger.

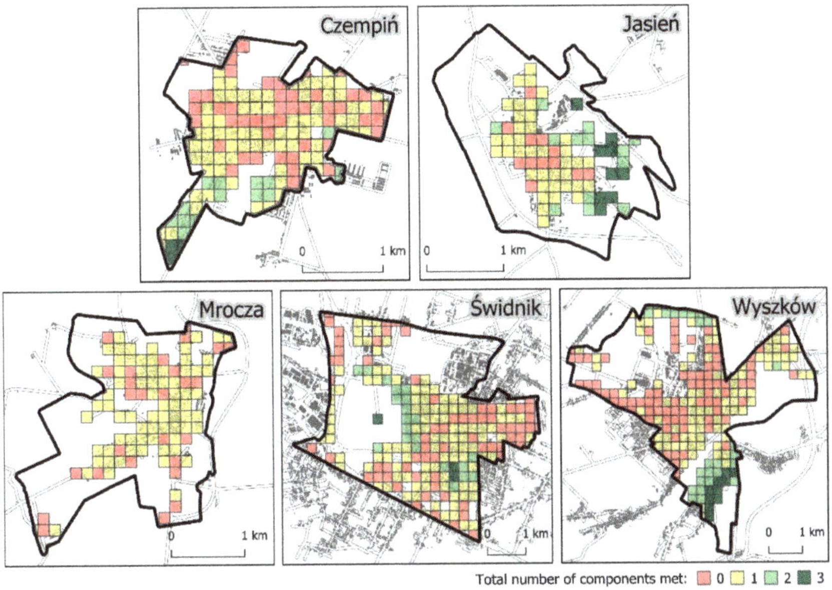

Figure 7. The results for the towns are displayed on the grid level. In this variant, the visibility condition was met if at least half of the observers in the building could see three or more trees and the building was within 300 m of a green area of 10 ha or larger.

4. Discussion

The 3-30-300 concept is a relatively young idea introduced by Konijnendijk in 2022 [40]. However, it has been scientifically justified, especially in relation to the "30-300" components [48]. In the case of the "3" component, many papers point to the benefits of visual access to green spaces [73]. One should bear in mind that the 3-30-300 idea is a general concept that does not include specific definitions regarding the qualitative features of green patches, such as their biodiversity, health conditions, or landscape/sight values. From one side, this flexibility enables the adaptation of the concept to the very specific characteristics of a town/city. As this is a relatively new concept, work is still underway to develop a methodology for calculating the indicators that it contains. Most research in this area, therefore, still consists of specific case studies with relatively small samples that differ in the methodology used. This can be somewhat of an obstacle both for scientific analysis and for common policies in terms of comparisons between cities or regions.

The aim of this article was, therefore, to propose an objective path for the future implementation of the 3-30-300 concept, which can be applied relatively easily with the available non-commercial ALS data [59] or with drone solutions that have recently become more common and competitive in many areas [74]. The issue has gained importance as some documents that recommend the incorporation of the 3-30-300 concept into urban spatial planning have recently been published [75]. Furthermore, there is a significant share of small and medium-sized towns among the urban areas of the EU. Approximately 95 million people live in 8,000 towns, constituting around 21% of the entire EU population. Two-thirds of these towns are densely populated, i.e., with 1500 inhabitants/km^2, with the total number of inhabitants ranging between 5000 and 50,000 [58]. We perceive the aforementioned facts as a foundation for justifying one of the subjects of this study, i.e., an exploration of the current state of adaptation to the 3-30-300 concept by smaller municipalities.

As stated by Hummel et al. [76] and Browning et. al. [48], the financial cost of ALS-based urban greenery inventories may be out of reach for some municipalities. However, ALS data might be useful for municipalities where tree surveys are not a priority, for instance, in urban domains connected with flood risk assessment, logistic network management, spatial planning, risk assessment, emergency management, advertising, research, promotion, and visualization. Therefore, a complex ALS solution can be a source of multidisciplinary data, which is a crucial element of a city management system.

The application of remote sensing tools and data to the realization of the 3-30-300 concept can theoretically provide spatially continuous information about every tree within an area of interest (AOI). This seems to be a great advantage over field-based inventories, as used by Croeser et al. [57], where only publicly accessible trees were surveyed (usually those along main streets). Furthermore, a larger number of green spaces requires the deployment of numerous surveyors on site, which incurs additional costs. In addition, field crew members vary in experience and may be prone to subjectivism. In contrast, the cost of objective RS-based methods does not increase linearly with the extent of an AOI [74]. Such an approach could help to further optimize tasks related to urban greenery inventories. Furthermore, the 3D character of RS data makes them an attractive material for publicity and visualization for future planning.

As a rule, most methods require successive optimization steps either before or during their implementation. The 3-30-300 rule is a general concept that offers many possibilities but also poses some challenges in practical implementation. From a technical point of view, an accurate assessment of the visibility component seems to be the most complex part of this method. Gathering and automatically classifying in situ images from each window seems most reliable [48] but is not practically feasible. Alternatively, one could conduct a survey, but it should always be analyzed with some confidence intervals regarding the response rates, their credibility, and their bias. In such situations, RS techniques could be very useful. As Croeser et al. [57] stated, three-dimensional spatial data (e.g., point clouds) could provide a robust material for visibility assessment in a single window. On the other hand, this would require enormous computational resources.

A more detailed analysis would require an examination of the viewing angle, i.e., the relation of a fixed/floating viewpoint to the window size, as well as the inclusion of any other facilities, such as balconies or terraces. Moreover, human eyes also have a main focus spot, prioritizing objects around the central part of the field of view over those situated at marginal regions. Therefore, to ensure the feasibility of the analysis, some aspects pertaining to visibility assessment had to be generalized in our study. Namely, there was only one observer for each side of the building, standing on the ground floor (1.7 m). In our approach, buildings as entities could meet the visibility requirement in the three variants described in the methodology. Again, this was just an arbitrary distinction originating from the lack of a precise definition of the visibility component. Therefore, this aspect should also be analyzed in more detail if the method is to be introduced at a larger scale. Nonetheless, the authors believe that the assumed simplifications could influence the results of the visibility assessment in larger cities, where there are more high-elevation buildings than in towns with low-rise buildings. All in all, the visibility component of the 3-30-30 method still requires further analysis regarding the definition of this component, as well as the development of specific algorithms that would facilitate its accurate determination.

Knowledge about the spatial distribution of trees can help to view the aspect of visibility from a reciprocal perspective. It is also possible to identify high-priority trees, i.e., trees that are seen by many observers (Figure 4). This type of information could be useful for planning revitalization measures or other urban infrastructure investments.

For example, our approach makes it possible to highlight weak spots, i.e., places where new planting is needed according to regulatory requirements and the actual expectations of residents.

Two main approaches are usually related to the determination of the "300" component: simple Euclidean distances and network analysis. The latter, which is relatively more accurate, is, at the same time, more complex. Proximity of a green area, e.g., a park, does not always ensure easy access. For instance, in the case of large fenced areas, the entrances constitute converging points. In order to determine the entry distance, detailed information about the spatial distribution of routes, pavement, junctions, etc., as well as their types and qualities, would be needed. On the other hand, one could reap the benefits of a park without accessing it just by being in its vicinity and/or by enjoying a view of it. This was the main reason why the simpler (Euclidean) approach was applied in this study. In our opinion, this might be a reliable proxy for a generic analysis and could be implemented in a greater number of municipalities, as few probably possess and would be willing to share detailed data on their communication networks.

Lastly, it seems paramount to emphasize that, currently, the 3-30-300 rule is a purely quantitative concept. It says little or nothing regarding the qualitative characteristics of green areas, which very often take precedence over pure numerical juxtapositions. Moreover, relying only on numerical summaries can lead to erroneous decisions. On this occasion, it is important to mention some qualitative features that could relatively easily expand the potential of the 3-30-300 method by using RS data. A complex spatial survey of the condition of trees could be carried out by calculating vegetation health indices such as the NDVI. Modern, relatively inexpensive unmanned aerial vehicles can be equipped with multispectral optical sensors that enable the creation of point clouds and increase the range of spectral information. This facilitates the ability to classify species, which could help determine biodiversity and differentiate subjective perceptions of the attractiveness of certain green spaces. This could also help identify and promote smaller or unknown places that are worth visiting.

Urban spatial management has become an important aspect for many cities. One of the biggest challenges is the provision and maintenance of high-quality green spaces. Other challenges include adapting to climate change and mitigating hazards resulting from extreme weather conditions, e.g., heavy rainfall, drought, and high temperatures. The United Nations' strategies and local conditions require municipal governments to develop strategies, policies, and programs to address the above issues [29]. For example, after 2024, all municipalities in Poland with at least 20,000 inhabitants will have to develop a local climate change adaptation strategy [77]. Therefore, in this article, the authors explored the possibilities of the methodological implementation of the 3-30-300 rule, which offers certain solutions for urban spatial management focused on the well-being of citizens. The use of a single remote sensing dataset leads to a relatively simple and standardized implementation of the proposed methodology in municipalities. It is all the more universal as the ALS data allow for analysis at flexible resolutions, the results of which could support decision-making chains at different scales (individual trees and buildings, streets, parks, neighborhoods, cities, agglomerations, and regions). Finally, with the proposed approach, it is relatively easy to mark the areas that are far from the minimum thresholds of the 3-30-300 rule. Such information can be helpful in objectively prioritizing tasks related to new initiatives and investments.

Local conditions play a crucial role in meeting the 3-30-300 provisions, e.g., when dense urban structures or unfavorable hydroclimatic conditions affect planting success and vegetation mortality rates [78]. The potential opportunities for shaping the 3-30-300 rule are also related to the size and density of urban infrastructures [79]. The growth of cities puts excessive pressure on existing infrastructure, making the creation and maintenance of existing green spaces an expensive and difficult task [80]. In order to overcome the aforementioned difficulties in fulfilling the criteria of the 3-30-300 rule in larger cities, several concepts have been presented by different authors. For example, Verheij et al. [81] point out the possibilities of ensuring public access to privately owned green spaces. Another approach described by Karteris et al. [80] utilizes the vertical dimension by proposing the implementation and development of green roofs.

It should also be considered that social acceptance of changes and investments can be an important factor. Therefore, measuring the actual use of green spaces and involving residents in the decision-making processes seems to be of utmost importance in a just society [82]. Such an approach proves invaluable when it comes to negotiations between different parties, such as decision makers, investors, and residents, where heated debates, often infused with subjective but important emotions, can be supported by impartial information.

5. Conclusions

The method presented in this study is a proposal of the adaptation of the 3-30-300 concept that demonstrates its practical applicability to the situation of smaller municipalities. In contrast to previous studies, the overriding premise of our method is based on the use of point clouds throughout the analytical process as the main source of data. The relatively simple implementation provides objective results that can be compared between towns, which, in turn, could be a valuable and objective decision-making tool. The method presented in this study was limited to a quantitative analysis based on the spatial distribution of green spaces. However, many aspects related to the assessment of qualitative features, as well as people's subjective perceptions, were discussed, and they certainly need to be taken into account in future adaptations.

Author Contributions: Conceptualization, M.C. and K.M.; methodology, K.M. and M.C.; data collection: K.G., S.K., T.H., K.M. and M.C.; validation, K.M., M.C., T.H., S.K., K.G. and K.K.; formal analysis, K.M. and M.C.; writing, K.M., M.C., T.H. and M.L.; visualization, K.M.; supervision, K.M. and M.C.; project administration, K.M. and M.C.; funding acquisition, M.C. All authors have read and agreed to the published version of the manuscript.

Funding: This study was carried out as part of the implementation of the "Strategic Consulting" project commissioned by the National Fund for Environmental Protection and Water Management as part of the "City with Climate" project.

Data Availability Statement: OpenStreetMap is an open data source licensed under the Open Data Commons Open Database License by the OpenStreetMap Foundation. The LiDAR data come from the Central Geodetic and Cartographic Resource in Poland. The data that support the findings of this study are available from the authors upon reasonable request.

Acknowledgments: We would like to thank Hieronim Kuśnierz, Aneta Modzelewska, Małgorzata Białczak, and Vasyl Mohytych for their support in this project. We would also like to thank the city representatives for their advice during the project.

Conflicts of Interest: The authors declare no conflicts of interest.

References

1. European Commission. *Directorate General for Research and Innovation. European Green Deal: Research & Innovation Call*; Publications Office: Luxembourg, 2021.
2. Dobrzańska, J.; Nadolny, A.; Kalbarczyk, R.; Ziemiańska, M. Urban Resilience and Residential Greenery—The Evidence from Poland. *Sustainability* **2022**, *14*, 11317. [CrossRef]

3. Surma, M.J. Green Infrastructure Planning as a Part of Sustainable Urban Development—Case Studies of Copenhagen and Wroclaw. *Proc. Latv. Univ. Agric. Landsc. Archit. Art* **2013**, *3*, 22–32.
4. Konijnendijk, C.C.; Ricard, R.M.; Kenney, A.; Randrup, T.B. Defining Urban Forestry—A Comparative Perspective of North America and Europe. *Urban For. Urban Green.* **2006**, *4*, 93–103. [CrossRef]
5. Aslanoğlu, R.; Kazak, J.K.; Szewrański, S.; Świąder, M.; Arciniegas, G.; Chrobak, G.; Jakóbiak, A.; Turhan, E. Ten Questions Concerning the Role of Urban Greenery in Shaping the Future of Urban Areas. *Build. Environ.* **2025**, *267*, 112154. [CrossRef]
6. Millennium Ecosystem Assessment. Ecosystems and Human Well-being: Synthesis. Island Press: Washington, DC, USA, 2005; ISBN 978-1-59726-040-4.
7. Evans, D.L.; Falagán, N.; Hardman, C.A.; Kourmpetli, S.; Liu, L.; Mead, B.R.; Davies, J.A.C. Ecosystem Service Delivery by Urban Agriculture and Green Infrastructure—A Systematic Review. *Ecosyst. Serv.* **2022**, *54*, 101405. [CrossRef]
8. Hegetschweiler, K.T.; Wartmann, F.M.; Dubernet, I.; Fischer, C.; Hunziker, M. Urban Forest Usage and Perception of Ecosystem Services—A Comparison between Teenagers and Adults. *Urban For. Urban Green.* **2022**, *74*, 127624. [CrossRef]
9. Nastran, M.; Pintar, M.; Železnikar, Š.; Cvejić, R. Stakeholders' Perceptions on the Role of Urban Green Infrastructure in Providing Ecosystem Services for Human Well-Being. *Land* **2022**, *11*, 299. [CrossRef]
10. Sun, R.; Chen, L. Effects of Green Space Dynamics on Urban Heat Islands: Mitigation and Diversification. *Ecosyst. Serv.* **2017**, *23*, 38–46. [CrossRef]
11. Liang, D.; Huang, G. Influence of Urban Tree Traits on Their Ecosystem Services: A Literature Review. *Land* **2023**, *12*, 1699. [CrossRef]
12. Pappalardo, V.; La Rosa, D.; Campisano, A.; La Greca, P. The Potential of Green Infrastructure Application in Urban Runoff Control for Land Use Planning: A Preliminary Evaluation from a Southern Italy Case Study. *Ecosyst. Serv.* **2017**, *26*, 345–354. [CrossRef]
13. Baró, F.; Gómez-Baggethun, E.; Haase, D. Ecosystem Service Bundles along the Urban-Rural Gradient: Insights for Landscape Planning and Management. *Ecosyst. Serv.* **2017**, *24*, 147–159. [CrossRef]
14. Karaczun, Z.M.; Michalak, W. *Wpływ Zmiany Klimatu i Zanieczyszczenia Powietrza na Zdrowie Mieszkańców Warszawy. Wyd*; PKEOM: Warszawa, Poland, 2019.
15. Rey-Gozalo, G.; Barrigón Morillas, J.M.; Montes González, D.; Vílchez-Gómez, R. Influence of Green Areas on the Urban Sound Environment. *Curr. Pollut. Rep.* **2023**, *9*, 746–759. [CrossRef]
16. Gołos, P. *Społeczne i Ekonomiczne Aspekty Pozaprodukcyjnych Funkcji Lasu i Gospodarki Leśnej—Wyniki Badań Opinii Społecznej*; Instytut Badawczy Leśnictwa: Sękocin Stary, Poland, 2018; ISBN 978-83-62830-68-8.
17. Fagerholm, N.; Eilola, S.; Arki, V. Outdoor Recreation and Nature's Contribution to Well-Being in a Pandemic Situation—Case Turku, Finland. *Urban For. Urban Green.* **2021**, *64*, 127257. [CrossRef] [PubMed]
18. Litleskare, S.; Calogiuri, G. Nature Visits during the COVID-19 Pandemic in Norway: Facilitators, Motives, and Associations with Sociodemographic Characteristics. *Front. Public Health* **2023**, *11*, 1138915. [CrossRef] [PubMed]
19. Brooks, S.K.; Webster, R.K.; Smith, L.E.; Woodland, L.; Wessely, S.; Greenberg, N.; Rubin, G.J. The Psychological Impact of Quarantine and How to Reduce It: Rapid Review of the Evidence. *Lancet* **2020**, *395*, 912–920. [CrossRef]
20. Derks, J.; Winkel, G.; Strieck, J.; De Vreese, R. Visitor Frequencies and Attitudes towards Urban Forests and Their Management, before and during the COVID-19 Lockdown. A Mixed Methods Case Study in Bonn, Germany. *Ecosyst. People* **2023**, *19*, 2195021. [CrossRef]
21. Raanaas, R.K.; Patil, G.G.; Hartig, T. Health Benefits of a View of Nature through the Window: A Quasi-Experimental Study of Patients in a Residential Rehabilitation Center. *Clin. Rehabil.* **2012**, *26*, 21–32. [CrossRef]
22. Dou, Y.; Zhen, L.; De Groot, R.; Du, B.; Yu, X. Assessing the Importance of Cultural Ecosystem Services in Urban Areas of Beijing Municipality. *Ecosyst. Serv.* **2017**, *24*, 79–90. [CrossRef]
23. Ciesielski, M.; Stereńczak, K. Using Flickr Data and Selected Environmental Characteristics to Analyse the Temporal and Spatial Distribution of Activities in Forest Areas. *For. Policy Econ.* **2021**, *129*, 102509. [CrossRef]
24. Mierzejewska, L.; Sikorska-Podyma, K.; Szejnfeld, M.; Wdowicka, M.; Modrzewski, B.; Lechowska, E. The Role of Greenery in Stress Reduction among City Residents during the COVID-19 Pandemic. *IJERPH* **2023**, *20*, 5832. [CrossRef]
25. Chen, Y.; Yue, W.; La Rosa, D. Which Communities Have Better Accessibility to Green Space? An Investigation into Environmental Inequality Using Big Data. *Landsc. Urban Plan.* **2020**, *204*, 103919. [CrossRef]
26. Elmqvist, T.; Gómez-Baggethun, E.; Langemeyer, J. Ecosystem Services Provided by Urban Green Infrastructure. In *Routledge Handbook of Ecosystem Services*; Routledge: London, UK, 2016; pp. 452–468.
27. Gałecka-Drozda, A.; Wilkaniec, A.; Szczepańska, M.; Świerk, D. Potential Nature-Based Solutions and Greenwashing to Generate Green Spaces: Developers' Claims versus Reality in New Housing Offers. *Urban For. Urban Green.* **2021**, *65*, 127345. [CrossRef]
28. Lafortezza, R.; Chen, J.; Van Den Bosch, C.K.; Randrup, T.B. Nature-Based Solutions for Resilient Landscapes and Cities. *Environ. Res.* **2018**, *165*, 431–441. [CrossRef]

29. United Nations General Assembly Transforming Our World: The 2030 Agenda for Sustainable Development 2015. Available online: https://sdgs.un.org/2030agenda (accessed on 22 December 2024).
30. Banaszak, K.; Gajda, M.; Hobot, A.; Mazur, M.; Renc, A.; Przyrodniczo-Klimatyczne Wskaźniki Zrównoważonego Rozwoju Miast. Przewodnik Dla Miast 2022. Available online: https://www.gov.pl/web/klimat/adaptacja-do-zmian-klimatu (accessed on 22 December 2024).
31. Browning, M.; Lee, K. Within What Distance Does "Greenness" Best Predict Physical Health? A Systematic Review of Articles with GIS Buffer Analyses across the Lifespan. *IJERPH* **2017**, *14*, 675. [CrossRef]
32. WHO. World Health Organization Urban Green Spaces: A Brief for Action. 2017. Available online: https://www.who.int/europe/publications/i/item/9789289052498 (accessed on 22 December 2024).
33. Pengelly Consulting. *"Nature Nearby": Accessible Natural Greenspace Guidance*; Natural England: York, UK, 2010.
34. Wysmułek, J.; Hełdak, M.; Kucher, A. The Analysis of Green Areas' Accessibility in Comparison with Statistical Data in Poland. *IJERPH* **2020**, *17*, 4492. [CrossRef]
35. McKernan, P.; Grose, M. An Analysis of Accessible Natural Greenspace Provision in the South East. *For. Comm.* **2007**. Available online: https://publications.naturalengland.org.uk/file/5620643968057344 (accessed on 22 December 2024).
36. Kabisch, N.; Strohbach, M.; Haase, D.; Kronenberg, J. Urban Green Space Availability in European Cities. *Ecol. Indic.* **2016**, *70*, 586–596. [CrossRef]
37. Cardinali, M.; Beenackers, M.A.; Van Timmeren, A.; Pottgiesser, U. The Relation between Proximity to and Characteristics of Green Spaces to Physical Activity and Health: A Multi-Dimensional Sensitivity Analysis in Four European Cities. *Environ. Res.* **2024**, *241*, 117605. [CrossRef]
38. Sejm of the Republic of Poland Obwieszczenie Marszałka Sejmu Rzeczypospolitej Polskiej z dnia 21 czerwca 2024 r. w sprawie ogłoszenia jednolitego tekstu ustawy o planowaniu i zagospodarowaniu przestrzennym. 2024. Available online: https://isap.sejm.gov.pl/isap.nsf/DocDetails.xsp?id=WDU20240001130 (accessed on 22 December 2024).
39. Hunter, A.J.; Luck, G.W. Defining and Measuring the Social-Ecological Quality of Urban Greenspace: A Semi-Systematic Review. *Urban Ecosyst.* **2015**, *18*, 1139–1163. [CrossRef]
40. Konijnendijk, C.C. Evidence-Based Guidelines for Greener, Healthier, More Resilient Neighbourhoods: Introducing the 3–30–300 Rule. *J. For. Res.* **2023**, *34*, 821–830. [CrossRef]
41. Koeser, A.; Hauer, R.; Andreu, M.; Northrop, R.; Clarke, M.; Diaz, J.; Hilbert, D.; Konijnendijk, C.; Landry, S.; Thompson, G.; et al. Using the 3-30-300 Rule to Assess Urban Forest Access and Preferences in Florida (United States). *Arboric. Urban For.* **2023**. [CrossRef]
42. Zhang, J.; Browning, M.H.E.M.; Liu, J.; Cheng, Y.; Zhao, B.; Dadvand, P. Is Indoor and Outdoor Greenery Associated with Fewer Depressive Symptoms during COVID-19 Lockdowns? A Mechanistic Study in Shanghai, China. *Build. Environ.* **2023**, *227*, 109799. [CrossRef]
43. Ling, K. A Look at Access to Green Space in Toronto Using the 3-30-300 Rule for Greener Cities. 2022. Available online: https://utoronto.scholaris.ca/items/97cff79a-09ad-4ede-b851-f85b8be60b27 (accessed on 22 December 2024).
44. Daland, S. An Investigation of the 3-30-300 Rule in a Swedish Context. 2023. Available online: https://gupea.ub.gu.se/bitstream/handle/2077/77317/B1232.pdf?sequence=1 (accessed on 22 December 2024).
45. Nieuwenhuijsen, M.J.; Dadvand, P.; Márquez, S.; Bartoll, X.; Barboza, E.P.; Cirach, M.; Borrell, C.; Zijlema, W.L. The Evaluation of the 3-30-300 Green Space Rule and Mental Health. *Environ. Res.* **2022**, *215*, 114387. [CrossRef] [PubMed]
46. Wyrzykowski, B.; Mościcka, A. Implementation of the 3-30-300 Green City Concept: Warsaw Case Study. *Appl. Sci.* **2024**, *14*, 10566. [CrossRef]
47. Battisti, L.; Aimar, F.; Giacco, G.; Devecchi, M. Urban Green Development and Resilient Cities: A First Insight into Urban Forest Planning in Italy. *Sustainability* **2023**, *15*, 12085. [CrossRef]
48. Browning, M.H.E.M.; Locke, D.H.; Konijnendijk, C.; Labib, S.M.; Rigolon, A.; Yeager, R.; Bardhan, M.; Berland, A.; Dadvand, P.; Helbich, M.; et al. Measuring the 3-30-300 Rule to Help Cities Meet Nature Access Thresholds. *Sci. Total Environ.* **2024**, *907*, 167739. [CrossRef]
49. Shahtahmassebi, A.R.; Li, C.; Fan, Y.; Wu, Y.; Lin, Y.; Gan, M.; Wang, K.; Malik, A.; Blackburn, G.A. Remote Sensing of Urban Green Spaces: A Review. *Urban For. Urban Green.* **2021**, *57*, 126946. [CrossRef]
50. Fekete, A.; Cserep, M. Tree Segmentation and Change Detection of Large Urban Areas Based on Airborne LiDAR. *Comput. Geosci.* **2021**, *156*, 104900. [CrossRef]
51. Kükenbrink, D.; Gardi, O.; Morsdorf, F.; Thürig, E.; Schellenberger, A.; Mathys, L. Above-Ground Biomass References for Urban Trees from Terrestrial Laser Scanning Data. *Ann. Bot.* **2021**, *128*, 709–724. [CrossRef]
52. Tang, L.; He, J.; Peng, W.; Huang, H.; Chen, C.; Yu, C. Assessing the Visibility of Urban Greenery Using MLS LiDAR Data. *Landsc. Urban Plan.* **2023**, *232*, 104662. [CrossRef]

53. Holopainen, M.; Vastaranta, M.; Hyyppä, J. Outlook for the Next Generation's Precision Forestry in Finland. *Forests* **2014**, *5*, 1682–1694. [CrossRef]
54. Zhang, K.; Hu, B. Individual Urban Tree Species Classification Using Very High Spatial Resolution Airborne Multi-Spectral Imagery Using Longitudinal Profiles. *Remote Sensing* **2012**, *4*, 1741–1757. [CrossRef]
55. Ciesielski, M.; Sterenczak, K. Accuracy of Determining Specific Parameters of the Urban Forest Using Remote Sensing. *iForest* **2019**, *12*, 498–510. [CrossRef]
56. Plowright, A.A.; Coops, N.C.; Eskelson, B.N.I.; Sheppard, S.R.J.; Aven, N.W. Assessing Urban Tree Condition Using Airborne Light Detection and Ranging. *Urban For. Urban Green.* **2016**, *19*, 140–150. [CrossRef]
57. Croeser, T.; Sharma, R.; Weisser, W.W.; Bekessy, S.A. Acute Canopy Deficits in Global Cities Exposed by the 3-30-300 Benchmark for Urban Nature. *Nat. Commun.* **2024**, *15*, 9333. [CrossRef]
58. Dijkstra, L.; Lecomte, L. *Towns in Europe: A Technical Paper*; DG REGIO Technical Papers; European Commission: Brussels, Belgium, 2023.
59. European Commission; Joint Research Centre; Florio, P.; Kakoulaki, G.; Martinez, A. *Non-Commercial Light Detection and Ranging (LiDAR) Data in Europe*; Publications Office: Luxembourg, 2021.
60. Kacperczyk, E.; Ciesielska, K.; Hernik, G.; Matysek-Zdun, U. *Powierzchnia i Ludność w Przekroju Terytorialnym w 2023 r./Area and Population in the Territorial Profile in 2023*; Główny Urząd Statystyczny (Statistics Poland): Warsaw, Poland, 2023.
61. Head Office of Geodesy and Cartography (GUGiK) Basic Information. 2024. Available online: https://www.gov.pl/web/gugik-en (accessed on 22 December 2024).
62. American Society for Photogrammetry and Remote Sensing (ASPRS) LAS Specification Version 1.2. Available online: https://www.asprs.org/a/society/committees/lidar/Downloads/ASPRS_LAS%201_2.pdf (accessed on 22 December 2024).
63. OpenStreetMap contributors Planet Dump. 2017. Available online: https://planet.osm.org (accessed on 22 December 2024).
64. *R Core Team R: A Language and Environment for Statistical Computing*; R Foundation for Statistical Computing: Vienna, Austria, 2023.
65. Roussel, J.-R.; Auty, D.; Coops, N.C.; Tompalski, P.; Goodbody, T.R.H.; Meador, A.S.; Bourdon, J.-F.; De Boissieu, F.; Achim, A. lidR: An R Package for Analysis of Airborne Laser Scanning (ALS) Data. *Remote Sens. Environ.* **2020**, *251*, 112061. [CrossRef]
66. Roussel, J.-R.; Auty, D. lidR: Airborne LiDAR Data Manipulation and Visualization for Forestry Applications 2016, 4.1.1. Available online: https://r-lidar.r-universe.dev/lidR (accessed on 22 December 2024).
67. Popescu, S.C.; Wynne, R.H. Seeing the Trees in the Forest. *Photogramm. Eng. Remote Sens.* **2004**, *70*, 589–604. [CrossRef]
68. Khosravipour, A.; Skidmore, A.K.; Isenburg, M.; Wang, T.; Hussin, Y.A. Generating Pit-Free Canopy Height Models from Airborne Lidar. *Photogramm. Eng. Remote Sens.* **2014**, *80*, 863–872. [CrossRef]
69. Pebesma, E. Simple Features for R: Standardized Support for Spatial Vector Data. *R J.* **2018**, *10*, 439. [CrossRef]
70. Pebesma, E.; Bivand, R. *Spatial Data Science: With Applications in R*, 1st ed.; Chapman and Hall/CRC: New York, NY, USA, 2023; ISBN 978-0-429-45901-6.
71. Padgham, M.; Lovelace, R.; Salmon, M.; Rudis, B. Osmdata. *JOSS* **2017**, *2*, 305. [CrossRef]
72. QGIS Development Team QGIS Geographic Information System; QGIS Association. Available online: http://www.qgis.org (accessed on 22 December 2024).
73. Labib, S.M.; Huck, J.J.; Lindley, S. Modelling and Mapping Eye-Level Greenness Visibility Exposure Using Multi-Source Data at High Spatial Resolutions. *Sci. Total Environ.* **2021**, *755*, 143050. [CrossRef] [PubMed]
74. Manfreda, S.; McCabe, M.F.; Miller, P.E.; Lucas, R.; Pajuelo Madrigal, V.; Mallinis, G.; Ben Dor, E.; Helman, D.; Estes, L.; Ciraolo, G.; et al. On the Use of Unmanned Aerial Systems for Environmental Monitoring. *Remote Sens.* **2018**, *10*, 641. [CrossRef]
75. United Nations Economic Commission for Europe Urban Forest Policy Brief; United Nations Economic Commission for Europe. 2023. Available online: https://land.unece.org/treesincities/ (accessed on 22 December 2024).
76. Hummel, S.; Hudak, A.; Uebler, E.H.; Falkowski, M.; Megown, K. A Comparison of Accuracy and Cost of LiDAR versus Stand Exam Data for Landscape Management on the Malheur National Forest. *J. For.* **2011**, *109*, 267–273. [CrossRef]
77. Sejm of the Republic of Poland Ustawa z Dnia 27 Listopada 2024 r. o Zmianie Ustawy—Prawo Ochrony Środowiska Oraz Niektórych Innych Ustaw 2024. Available online: https://isap.sejm.gov.pl/isap.nsf/DocDetails.xsp?id=WDU20240001940 (accessed on 22 December 2024).
78. Pataki, D.E.; Alberti, M.; Cadenasso, M.L.; Felson, A.J.; McDonnell, M.J.; Pincetl, S.; Pouyat, R.V.; Setälä, H.; Whitlow, T.H. The Benefits and Limits of Urban Tree Planting for Environmental and Human Health. *Front. Ecol. Evol.* **2021**, *9*, 603757. [CrossRef]
79. McDonald, R.I.; Aronson, M.F.J.; Beatley, T.; Beller, E.; Bazo, M.; Grossinger, R.; Jessup, K.; Mansur, A.V.; Puppim De Oliveira, J.A.; Panlasigui, S.; et al. Denser and Greener Cities: Green Interventions to Achieve Both Urban Density and Nature. *People Nat.* **2023**, *5*, 84–102. [CrossRef]
80. Karteris, M.; Theodoridou, I.; Mallinis, G.; Tsiros, E.; Karteris, A. Towards a Green Sustainable Strategy for Mediterranean Cities: Assessing the Benefits of Large-Scale Green Roofs Implementation in Thessaloniki, Northern Greece, Using Environmental Modelling, GIS and Very High Spatial Resolution Remote Sensing Data. *Renew. Sustain. Energy Rev.* **2016**, *58*, 510–525. [CrossRef]

81. Verheij, J.; Ay, D.; Gerber, J.-D.; Nahrath, S. Ensuring Public Access to Green Spaces in Urban Densification: The Role of Planning and Property Rights. *Plan. Theory Pract.* **2023**, *24*, 342–365. [CrossRef]
82. Rigolon, A. Nonprofits and Park Equity in Los Angeles: A Promising Way Forward for Environmental Justice. *Urban Geogr.* **2019**, *40*, 984–1009. [CrossRef]

land

MDPI

Article

Connecting Natural and Planted Forests: New Ecological Functions in an Agricultural Landscape in Northern Spain

Javier Brazuelo Núñez [1], Carlos A. Rivas [2], Guillermo Palacios-Rodríguez [2] and Rafael M. Navarro-Cerrillo [1,*]

1 Laboratory of Dendrochronology, Silviculture and Global Change—Den-drodatLab—ERSAF, Department of Forest Engineering, University of Cordoba, Campus de Rabanales, Crta. IV, km. 396, 14071 Córdoba, Spain; j.brazuelo@alumnos.upm.es

2 Mediterranean Forest Global Change Observatory, Digitalization and Development in Forestry Ecosystems Laboratory, DigiFoR+-ERSAF, Department of Forestry Engineering, University of Cordoba, Campus de Rabanales, Crta. IV km. 396, 14071 Córdoba, Spain; z82ricoc@uco.es (C.A.R.); gpalacios@uco.es (G.P.-R.)

* Correspondence: rmnavarro@uco.es

Abstract: The connectivity of forest ecosystems is increasingly recognized as a key factor in evaluating the sustainability of forest management, with significant implications for biodiversity conservation. This study examines the impact of afforestation programs on forest evolution, fragmentation, and connectivity in León province, Spain, over the past 25 years (1996–2020). Three scenarios were modeled across two periods (1996–2006 and 2006–2020), integrating data from the national forest inventories (IFN2, IFN3, and IFN4) and afforestation program records provided by the Junta de Castilla y León. The evolution of connectivity "with" and "without" afforestation was analyzed using Graphab 2.6 and graph theory, and several connectivity metrics were calculated. The first period analyzed, influenced by the two initial afforestation programs, corresponded to the end of a forest expansion phase, followed by a decrease in tree cover. Despite this reduction, a net positive balance of up to 24% of all connectivity metrics (NC, PC, Flux, and ECA) was observed throughout the study period. Afforestation in mountain areas enhanced tree cover continuity, resulting in a more homogeneous but less diverse landscape. Conversely, afforestation in agricultural lands increased landscape heterogeneity, diversifying and extending the ecological network of connections. These programs have played a crucial role in shaping the landscape, influencing its diversity and the evolution of forest connectivity. Legislation grounded in technical and ecological principles should be prioritized as a strategic tool to address pressing land management challenges and preserve natural values.

Keywords: afforestation; connectivity; fragmentation; land use change; landscape ecology

1. Introduction

The ongoing global biodiversity crisis, driven mainly by human activities, affects ecosystems across the planet, including in the Iberian Peninsula. The expansion of human populations has led to biotic homogenization, resource overexploitation, and an increased susceptibility of ecosystems to both biotic and abiotic disturbances, threatening the preservation of natural habitats [1]. Added to this is climate change, driven by the consequences of anthropogenic expansion, which makes it necessary to incorporate new technical and ecological criteria into land management and planning [2].

Afforestation may bring beneficial changes to the landscape; however, to evaluate its environmental costs and benefits, it is important to understand the link between afforestation programs and landscape functionality. In Europe, the afforestation of agricultural lands has been promoted as a key tool for environmental management. In Spain, over 876,000 hectares have been reforested since 1993 under regulations like CEE-2080/92 [3]. Castilla y León, particularly the province of León, is the region with the largest reforested area, owing to programs such as the Regional Afforestation Program for Agricultural Lands. Some of these afforestation policies have altered fragmentation patterns, increased connectivity, and reduced distances between forest areas with different land covers [4]. Landscape fragmentation, i.e., the division of a habitat into smaller patches, affects ecosystem cycles, biodiversity, and species movement [5]. This phenomenon, which is particularly affecting Mediterranean landscapes, has been exacerbated by factors such as the abandonment of traditional practices, agricultural intensification, and forest management.

Forest connectivity is a key aspect for assessing the sustainability of forest management and studying changes in forest landscapes [6,7]. Connectivity facilitates species movement, genetic exchange, and other essential ecological flows [8]. Therefore, it is a fundamental factor to consider in various fields, such as ecological restoration, the delimitation of new protected areas, invasive species control, and the management of transboundary resources [8–10]. Connectivity is a fundamental landscape quality for counteracting the negative effects of habitat loss and fragmentation [11,12] and, as such, must be evaluated at the landscape level [13]. However, forest expansion can have negative effects, and increased connectivity may have detrimental consequences [14].

Our study addresses the need to evaluate the ecological consequences of afforestation programs, focusing on the functional connectivity of forest landscapes in the province of León. This region, characterized by its contrasting relief between the Central Plateau and the Cantabrian Mountains, hosts great biological diversity and forest formations, including natural forests and recent afforestation. Using graph theory, we analyzed land cover dynamics and their impact on functional ecological connectivity, with a particular focus on medium-sized carnivorous forest mammals as the reference species [15,16]. This approach, which has been proven effective for modeling landscapes at different scales [17], helps to understand how these transformations affect connectivity in a context marked by rural depopulation, the abandonment of traditional practices, and climate change. The main objective was to evaluate, both qualitatively and quantitatively, the contribution of the EU afforestation program to the functional ecological connectivity of forests in central Spain. To achieve this, we analyzed the dynamics of land cover changes in the study area, focusing on forested areas, and evaluated the progression of ecological connectivity, considering the physical differences in landscape configuration. The implementation of well-designed afforestation programs can serve as a pivotal strategy to enhance functional ecological connectivity in forested landscapes, fostering biodiversity conservation, ecosystem resilience, and sustainable land management.

2. Materials and Methods

2.1. Study Area

The study was conducted in the province of León, located in the northwest of the Autonomous Community of Castilla y León. The province is characterized by a diverse landscape, with mountain areas (medium and high elevation), moors, riparian systems, and large agricultural areas. The region has an average slope of over 13% and an average altitude of more than 1000 m above sea level. It is a transitional area between the Mediterranean and Euro-Siberian biogeographic regions (Figure 1).

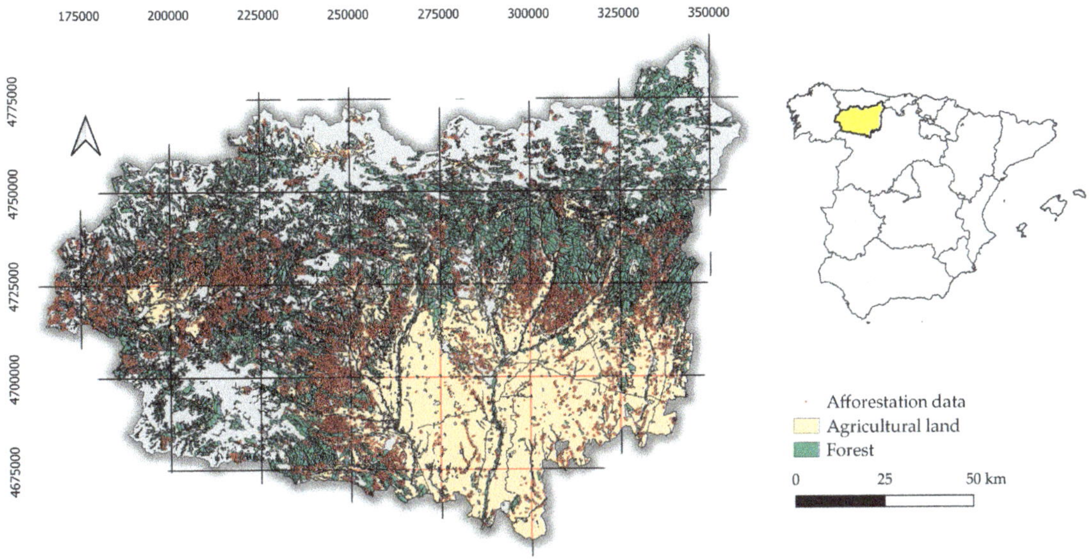

Figure 1. Representation and location of the study area in Spain (yellow), showing forests, agricultural lands, and afforestation data.

2.2. Land-Use Data Sets and Scenarios

The source of basic cartographic information used was the time series of the Forest Map of Spain (MFE): (1) Forest Map of Spain Scale 1:200,000 (MFE200 Ruiz de la Torre), cartographic base of the second National Forest Inventory (IFN2, 1987–1996), digitized at a scale of 1:50,000. (2) Forest Map of Spain Scale 1:50,000 (MFE50. Ministry of Agriculture, Food and Environment. General Directorate of Rural Development and Forest Policy), cartographic base of the third National Forest Inventory (IFN3, 1997–2006). (3) Forest Map of Spain Scale 1:25,000 (MFE25. Ministry for Ecological Transition and the Demographic Challenge. General Directorate of Biodiversity, Forests and Desertification), cartographic base of the fourth National Forest Inventory (IFN4, 2007–2020). The CORINE Land Cover 1990 map was also used. This map was chosen because the MFE200 did not include land use data. Therefore, the forested areas from the MFE200 were combined with the other land use types from CORINE Land Cover 1990 to create a single map.

Since the layers used contained multiple land covers, they were reclassified or grouped into 12 land use categories: agriculture, artificial areas, water bodies, shrubland, bare ground, grasslands and pastures, and forested areas. The forested areas were further reclassified into three types: coniferous forests, broadleaf forests, and mixed coniferous and broadleaf forests. Subsequently, afforestation was added as follows: afforestation of coniferous forests, afforestation of broadleaf forests, and afforestation of mixed forests (Table 1).

Table 1. Classification, description, and costs of the coverage systems grouped in the study area.

Land Cover	Definition	Cost
Agriculture	Agricultural lands. This includes trees scattered among crops and treeless mosaics.	20
Artificial	Urban space and discontinuous urban mosaics, linear communication infrastructures and spaces dedicated to mining, and waste dumps and landfills.	100

Table 1. *Cont.*

Land Cover	Definition	Cost
Coniferous forests	Pine forests and mixtures of native conifers and other allochthonous species. Juniper and juniper forests are included.	1
Broadleaf forests	Beech, birch, Spanish oak, hazel, ash, and other riparian forests. Holm oak and cork oak forests. Eucalyptus and other high-yield species plantations.	1
Mixed coniferous and broadleaf forests	Mixture of native and non-native conifers and broadleaf trees in the Mediterranean and Atlantic regions.	1
Water bodies	Swamps, reservoirs, wetlands, and high mountain lagoons.	100
Shrubland	Bush and shrub formations	5
Bare ground	Mountain without upper vegetation. This includes burned areas and deforested forests due to the felling of production species, bare soils and rocky outcrops.	5
Grasslands and pastures	Grasslands and meadows. This includes mixtures of grasslands and shrublands, and grasslands with scattered shrubs.	10
Afforestation of coniferous forests	Afforestation included in the different community programs with dominance (90% or more) of coniferous species.	1 (5) *
Afforestation of broadleaf forests	Afforestation included in the different community programs with dominance (90% or more) of broadleaf species.	1 (5) *
Afforestation of mixed forests	Afforestation included in the different community programs with dominance (less than 90%) of coniferous or broadleaf species.	1 (5) *

* Afforestation data have a cost value of 1 (habitat) for scenarios 0, 1 and 2, and a value of 5 for the hypotheses of zero afforestation (Assumptions 1, 2a, and 2b).

The Junta de Castilla y León provided the afforestation data. According to the year of planting, these data corresponded to different afforestation programs in Castilla y León. Afforestation data were classified according to their composition: (1) broadleaf, (2) conifers, and (3) mixed afforestation, with less than 90% dominance of one of these two types. Three scenarios were defined, each one characterized by a cover system.

- Scenario 0: The MFE200 with the incorporation of the artificial coverage of the CORINE Land Cover 1990. This scenario represents the initial situation, before the afforestation programs.
- Scenario 1: Intermediate or transition scenario, consisting of the MFE50 cartographic base with the first two afforestation programs: (i) program 1993–1999: Regional Program for Reforestation of Agricultural Lands-Regulation (CEE) 2080/92 and Royal Decree 378/93 (for the study area, 23,369 afforested stands with a total of 32,048 hectares, average size of 1.37 ha); (ii) program 2000–2006: Regulation 1257/99 and Royal Decrees 6/2001 and 708/2002 (15,479 afforested stands with a total of 18,225 ha, an average size of 1.17 ha).
- Scenario 2: a final scenario, consisting of the cartographic base of the MFE25 and the afforestation data of the (i) third program 2007–2013: Regulation (EC) No 1698/2005 (9041 afforested stands with a total of 10,269 ha, average size of 1.13 ha); (ii) fourth program 2014–2020: Regulation (EU) 1305/2013 (1111 afforested stands with a total of 1362 ha, average size of 1.22 ha).

2.3. *Dynamics of Land Changes*

Changes in cover were quantified by crossing the different scenarios using GIS tools. This was performed in two hops, for the changes between the initial and transition scenarios

(scenarios 0–1) and between transition one and the final scenario (scenarios 1–2). The balance was expressed using a Sankey diagram. The area and proportion of the afforestation were also calculated.

2.4. Fragmentation Metrics and Functional Connectivity

For the fragmentation and functional connectivity calculations, the vector layers were rasterized to a 25 × 25 m raster because the connectivity software operated with raster (described later in this section) and with the aim to homogenize the different temporal layers. Two fragmentation metrics were calculated for the assumptions and hypotheses: mean patch size and number of patches [18]. These are the basic metrics to measure fragmentation and are necessary to understand and analyze the connectivity results.

Functional connectivity was assessed using graph theory. Graphs are mathematical structures composed of a set of nodes and links that represent the entire forest landscape, such that each patch or distinct unit of forest is represented by a node. Two nodes may or may not be functionally connected by a link, allowing for direct dispersion between that pair of nodes [19]. The techniques for evaluating functional connectivity in a landscape are based on one or more reference species. Medium and large mammals are the most widely used faunal group due to their "umbrella species" status [20,21]. The study selected a profile of medium-sized carnivorous mammals, using the Badger (*Meles meles*), Wildcat (*Felis silvestris*), and Marten (*Martes foina*) as reference species. These mammals use forests as their primary habitat, have regional movement ranges, are widely distributed in the study area, and are particularly sensitive to changes in the landscape [22–24].

Functional connectivity was calculated using Graphab 2.6 software (http://thema.univ-fcomte.fr/productions/graphab/ accessed on 29 on August 2022). To model the network of connections between habitat patches, a resistance raster map was used, considering forest areas as "habitat" (Table 1). Subsequently, links between habitats were created. The links were calculated as the minimum cost distance between nodes, with a 5 km dispersal distance defined by the dispersal characteristics of the reference faunal group. To calculate functional connectivity, Euclidean distance was transformed into cost distance. The dispersal distance metric was adjusted to a cost distance of 2718.78 from the set of trajectories using linear regression [25,26]. The costs represent the permeability of the terrain, where high costs correspond to low-permeability areas, and low costs correspond to highly permeable areas that facilitate connectivity [25]. Once the links were calculated, connectivity metrics were computed. Both global and local metrics were selected. Global metrics measure the functional connectivity of the entire landscape and are useful for evaluating landscape evolution and measuring the global consequences of a particular action. The global metrics calculated were the Number of Components (NC), Probability of Connectivity (PC), Flux (F), and Equivalent Connected Area (ECA). Local metrics describe connectivity for each element and allow us to characterize the importance of the patches and links; here, the metrics calculated were Flux (F) and Current Flow (CF) (Table 2).

After calculating the three scenarios (0, 1, and 2), the fragmentation and functional connectivity analysis was reformulated by replacing the afforested tiles with others that, while not an impediment to the flow or movement of species, did not constitute habitat. Three hypotheses were made: Assumption 1: a null hypothesis of the occurrence of afforestation programs (1993–2006). Assumption 2a: a null hypothesis of the occurrence of afforestation programs (2007–2020). Assumption 2b: a null hypothesis of the occurrence of afforestation programs (1993–2020) (Figure 2).

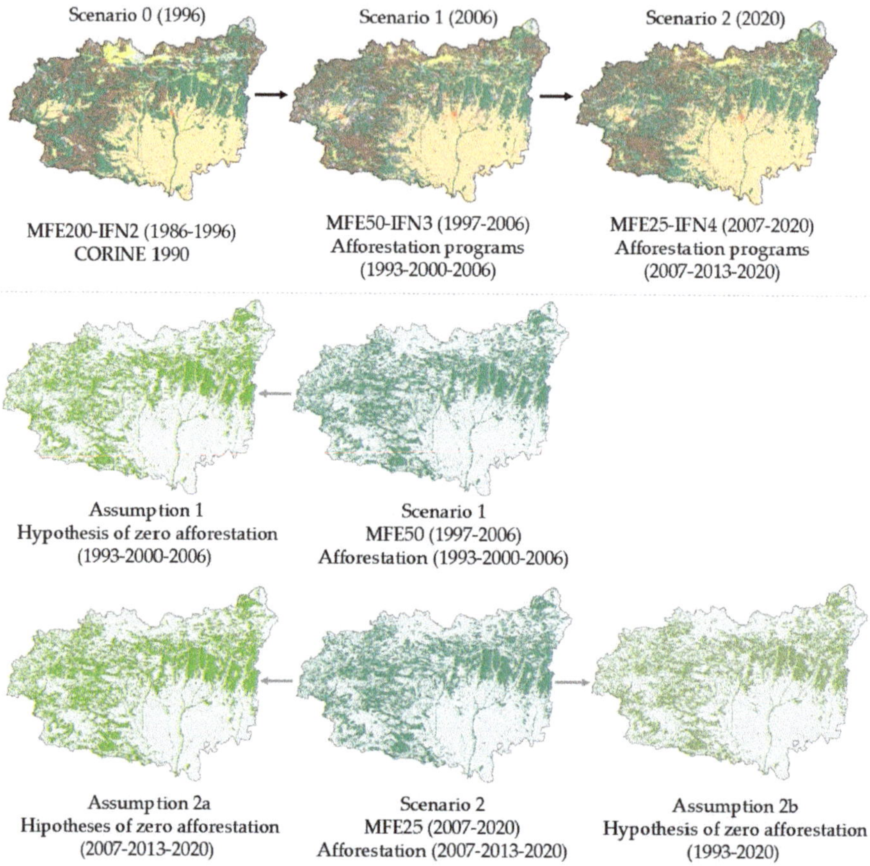

Figure 2. Representation of the temporal dynamics of the study, expressed through the scenarios and assumption.

Table 2. Description of connectivity metrics analyzed according to Clauzel et al. (2019) [27]. Ecological meaning according to bibliographic references and Rivas et al. (2024a, b) [15,16].

Flux (F)	Formula	Meaning	Ecological Meaning	Reference
Global level	$S\#F = \sum_{i=1}^{n} \sum_{\substack{j=1 \\ J \neq i}}^{n} a_j^{\beta}\, e^{-ad_{ij}}$	Sum of potential dispersion from all patches	The importance of the F metric reflects the importance of the patch network, with a higher F metric meaning greater connectivity between patches, facilitating the ecological factors that depend on connectivity	[26,28,29]
Local level	$\mathrm{Fi} = \sum_{\substack{j=1 \\ J \neq i}}^{n} a_j^{\beta} e^{-ad_{ij}}$	Sum of capacity of patches other than i and weighted according to their minimum distance to the focal patch through the graph. This sum is an indicator of the potential dispersion. from patch i, or conversely, to patch i.		

Table 2. *Cont.*

Flux (F)	**Formula**	**Meaning**	**Ecological Meaning**	**Reference**
Equivalent Probability (ECA)	Formula	Meaning	Ecological Meaning	Reference
Global level	$EC = \sqrt{\sum_{i=1}^{n} \sum_{j=1}^{n} a_i\, a_j\, e^{-ad_{ij}}}$	Square root of the sum of products of capacity of all pairs of patches weighted by their interaction probability	ECA allows us to interpret the dynamics in functional connectivity in relation to the different types of changes in habitat area and to assess to what extent the area gains or losses are actually beneficial or detrimental to maintaining and promoting ecological flows throughout the landscape.	[30]
Probability of Connectivity (PC)	Formula	Meaning	Ecological Meaning	Reference
Global level	$PC = \frac{1}{A^2} \sum_{i=1}^{n} \sum_{j=1}^{n} a_i\, a_j\, e^{-ad_{ij}}$	Sum of products of capacity of all pairs of patches weighted by their interaction probability, divided by the square of the area of the study zone. This ratio is the equivalent to the probability that two points randomly placed in the study area are connected.	The probability of the connectivity index (PC) is defined as the probability that two animals randomly placed in the landscape fall into habitat areas that are reachable from each other (interconnected).	[19]
Number of Components (NC)	Formula	Meaning	Ecological Meaning	Reference
Global level	NC = nc	Number of components of the graph.	A component is a group of connected nodes, meaning that organisms can move (link) between patches (nodes) into the same component, and therefore, patches of different components cannot communicate because they are isolated. This implies that within the same component, there is genetic exchange or other ecological functions on which connectivity depends, but this exchange does not exist between the different components.	[28]

Table 2. *Cont.*

Flux (F)	Formula	Meaning	Ecological Meaning	Reference
Current Flow (CF)	Formula	Meaning	Ecological Meaning	Reference
Local level	$CF = \sum_{j}^{n} a_i^j$	Sum of currents passing through patch i. C_i^j represents the current through patch i when currents are sent from all patches (except j) to patch j. Patch j is connected to the ground.	The graphical structure and the calculation of the weighted dispersion flux reflect the importance of the connectivity of the habitat patches; therefore, patches with higher metrics indicate greater connectivity.	[31]

N: number of patches. nc: Number of components. a_i: Capacity of patch i (generally the surface area). A: Area of the study zone. d_{ij}: Distance between patches i and j (generally the least-cost distance between them). $e^{-\alpha dij}$: Probability of movement between patches i and j. α = Brake on movement distance. β = Exponent-to-weight more or less capacity.

2.5. Statistical Analysis

A Student's *t*-test of variance was applied for comparing the connectivity metrics between the scenarios (scenario 0, 1996; scenario 1, 2006; scenario 2, 2020) and assumptions (Assumption 1: no afforestation, 1993–2006; Assumption 2a: no afforestation 2007–2020; and Assumption 2b: no afforestation 1993–2020). Levene's test was used to check the equality of variance.

3. Results

3.1. Dynamics of Land Changes

In the first period (scenarios 0–1, between 1996 and 2006), 35% of the study area experienced some type of change, while in the second period (scenarios 1–2, between 2006 and 2020), more than 95% remained unchanged. During the first period, significant changes and transitions occurred among broadleaf forests, shrublands, and agricultural lands. In the second period, most of the changes were attributed to afforestation data (Table 3 (3.a) and Figure 3). The afforestation programs collectively covered 62,000 hectares (4.1% of the territory). In the first two programs, the reforested area was five times larger than in the last two. In both cases, afforestation was dominated by coniferous trees, followed by mixed forests and, finally, broadleaf forests (Table 3 (3.b)). The first two programs were implemented primarily on shrubland cover, followed by other forest units and agricultural lands, whereas the last two programs were carried out mainly on agricultural lands, followed by shrublands and other forest covers (Table 3 (3.a)).

Table 3. Land use changes and the different afforestation efforts carried out in the study area. Table 3.a: Proportion relative to the study area and variation (%) of the main land covers; 3.b: area and proportion of the specific composition of the afforestation effort.

3.a	Scenario 0 (1996)		Scenario 1 (2006)		Scenario 2 (2020)		Period 1 Scenarios 0–1	Period 2 Scenarios 1–2
Land use	Area (ha)	%AE	Area (ha)	%AE	Area (ha)	%AE	Variation in %	
Agriculture	475.273	30.5	514.316	33.0	517.889	33.2	8.2	0.7
Artificial	13.663	0.9	28.726	1.8	28.696	1.8	110.2	−0.1
Water bodies	5.802	0.4	7.197	0.5	7.209	0.5	24.0	0.2
Shrubland	397.467	25.5	364.515	23.4	369.623	23.7	−8.3	1.4

Table 3. *Cont.*

3.a	Scenario 0 (1996)		Scenario 1 (2006)		Scenario 2 (2020)		Period 1 Scenarios 0–1	Period 2 Scenarios 1–2
Bare ground	53.553	3.4	19.656	1.3	23.736	1.5	−63.3	20.8
Grasslands and pastures	97.478	6.3	74.773	4.8	75.972	4.9	−23.3	1.6
Coniferous forests	102.448	6.6	133.704	8.6	133.360	8.6	30.5	−0.3
Broadleaf forests	399.489	25.6	383.986	24.6	385.087	24.7	−3.9	0.3
Mixed forests	13.611	0.9	32.333	2.1	17.149	1.1	137.6	−47.0
Forest	515.548	33.1	550.023	35.3	535.596	34.4	6.7	−2.6

3.b	Scenario 1 (Afforestation 1993–2006)		Scenario 2 (Afforestation 2007–2020)		Total afforestation programs	
Composition	Area (ha)	%	Area (ha)	%	Area (ha)	%
Coniferous	26.493	52.7	7.205	61.9	33.698	54.4
Broadleaf	3.369	6.7	473	4.1	3.842	6.2
Mixed forests	20.454	40.7	3.963	34.0	24.417	39.4
Total	50.316		11.641		61.957	

Proportion %AE refers to the percentage of each cover with respect to the total study area in each scenario. Variation in % refers to the increase or decrease for each cover in one scenario with respect to the previous one.

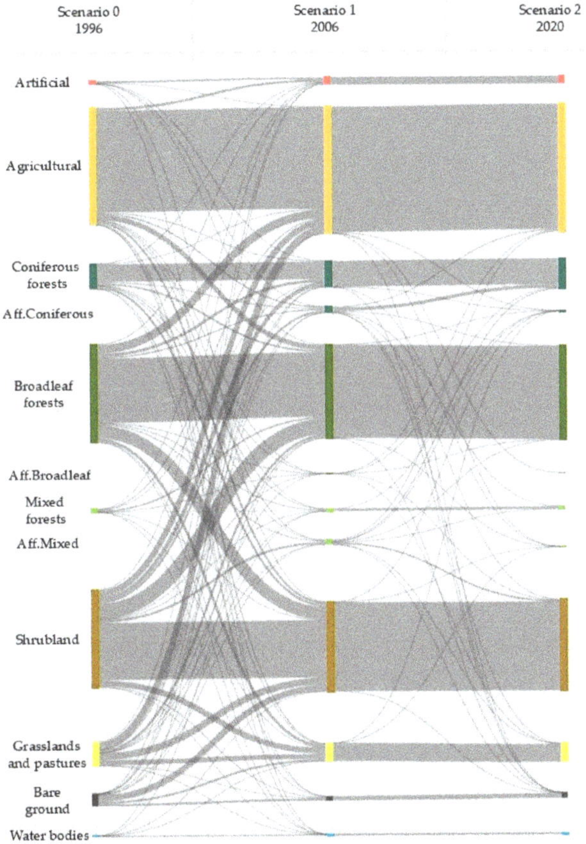

Figure 3. Sankey matrix. Representation of the dynamics and succession of changes in land use and cover in the study area across the three scenarios.

3.2. Fragmentation and Connectivity Analyses

Regarding the configuration of the network of connections, the three scenarios generated a similar landscape in terms of distribution and structure. Each scenario featured a large expanse of habitat in the northern region extending from east to west, characterized by the presence of large patches. This area corresponded to the mountain region in the study zone. The southeastern quadrant exhibited very low habitat density and patch size, while the southwestern quadrant displayed a mixture of the two previous patterns, alternating between areas devoid of habitat and others densely occupied by forested zones. The southeastern quadrant was distinguished by a central axis of habitat and a north-south movement pattern. The initial scenario depicted a simple system, with medium-sized habitat patches and a small total area. However, scenarios 1 and 2 exhibited more complex systems, particularly in the southern half, with a significant increase in both patch density and connectivity (Figure 4).

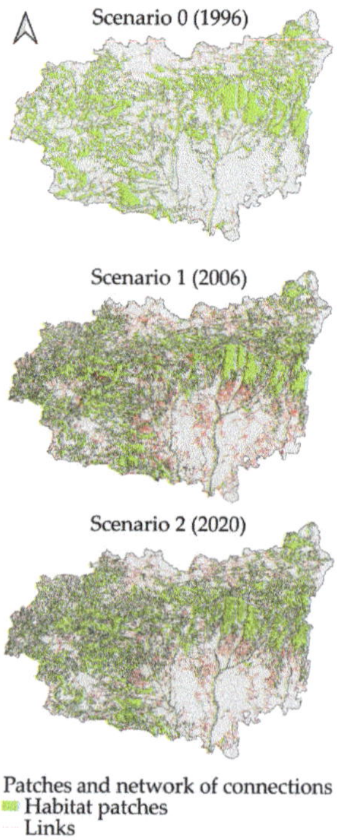

Figure 4. Evolution of the links between patches in the different scenarios. Forest patches are shown in dark green and links between forest patches are shown in red.

The overall trend in the landscape initially showed an increase in fragmentation, with a rise in patch number and a decrease in patch size (from scenario 0 to 1). This trend was reversed (from scenario 1 to 2) with an increase in patch size and a decrease in the number of patches. The analysis of the effect of afforestation on the landscape showed two evident trends. In scenario 1 and Assumption 1, afforestation efforts reduced fragmentation by increasing patch size and decreasing the number of patches. This pattern was also observed between scenario 2 and scenario 2a. However, an opposite trend was observed between scenario 2 and Assumption 2b, with an increase in fragmentation (Table 4).

Table 4. Evolution of fragmentation: Number of patches, average size, and total areas in hectares for the different scenarios.

Scenario	Number of Patches	Average Size (ha)	Total (ha)
Scenario 0 (1996)	415	1242	515,534
Scenario 1 (2006)	7288	75	549,899
Assumption 1 (No afforestation 1993–2006)	9432	52	499,689
Scenario 2 (2020)	5954	90	535,419
Assumption 2a (No afforestation 2007–2020)	5199	101	523,937
Assumption 2b (No afforestation 1993–2020)	9948	50	496,166

The final scenario exhibited significantly higher values than the initial scenario (e.g., 24% in ECA and 55% in PC). The final scenario (scenario 2) showed slightly lower values than the intermediate scenario, especially for the Flux metric. In the case of the ECA, a 3% decrease was observed, while the Probability of Connectivity (PC) showed a 5% loss in the final scenario. As for the number of components, the final scenario presented four times more components than the intermediate scenario, with eight and two components, respectively. In this case, scenario 0 had an intermediate situation, with four components (Table 5).

Table 5. Evolution of global connectivity metrics in the study periods: Number of Components (NC), Probability of Connectivity (PC), Flux (F), and Equivalent Connected Area (ECA) for the different scenarios.

Scenario	NC	PC	Flux	ECA
Scenario 0 (1996)	4	0.022614	8.41×10^{11}	3.88×10^9
Scenario 1 (2006)	2	0.037332	2.44×10^{13}	4.99×10^9
Assumption 1 (No afforestation 1993–2006)	5	0.026081	2.56×10^{13}	4.17×10^9
Scenario 2 (2020)	8	0.035216	1.94×10^{13}	4.84×10^9
Assumption 2a (No afforestation 2007–2020)	6	0.033530	1.72×10^{13}	4.72×10^9
Assumption 2b (No afforestation 1993–2020)	6	0.027221	2.81×10^{13}	4.25×10^9

In the afforestation programs from 1993 to 2006 (Assumption 1), the global connectivity metrics showed that in the scenario without afforestation, there was a 30% decrease in PC and a 15% decrease in ECA. However, the global Flux exhibited a slight (5%) increase. The number of components increased from two to five units (Table 4), primarily located in the agricultural areas of the southeastern quadrant of the study area (Figure 4).

For the local metrics (Flux), Assumption 1 without afforestation showed a decrease in the accumulated central values (25–75 quartiles), as well as a drop of about 20% in the mean and maximum values (Figures 5 and 6). This reduction in the metric values was common throughout the study area. In the areas with the greatest density of habitat patches, an increase in the number of patches was observed in the assumption without afforestation. However, in areas with a low habitat presence, the opposite effect occurred, with a decrease in the number of patches, but with the same final result: a decline in the Flux values (Figure 6).

Current Flow (CF) showed a higher number of receptor patches for large corridors in the assumption without afforestation. In scenario 1, there was a single large receptor patch, with its centroid located to the west of the center of the study area, and with a higher potential as a corridor connecting eastward and southeast. In Assumption 1, this central patch was divided into two, with centroids at the center and with strong horizontal and southeast flows (Figure 7).

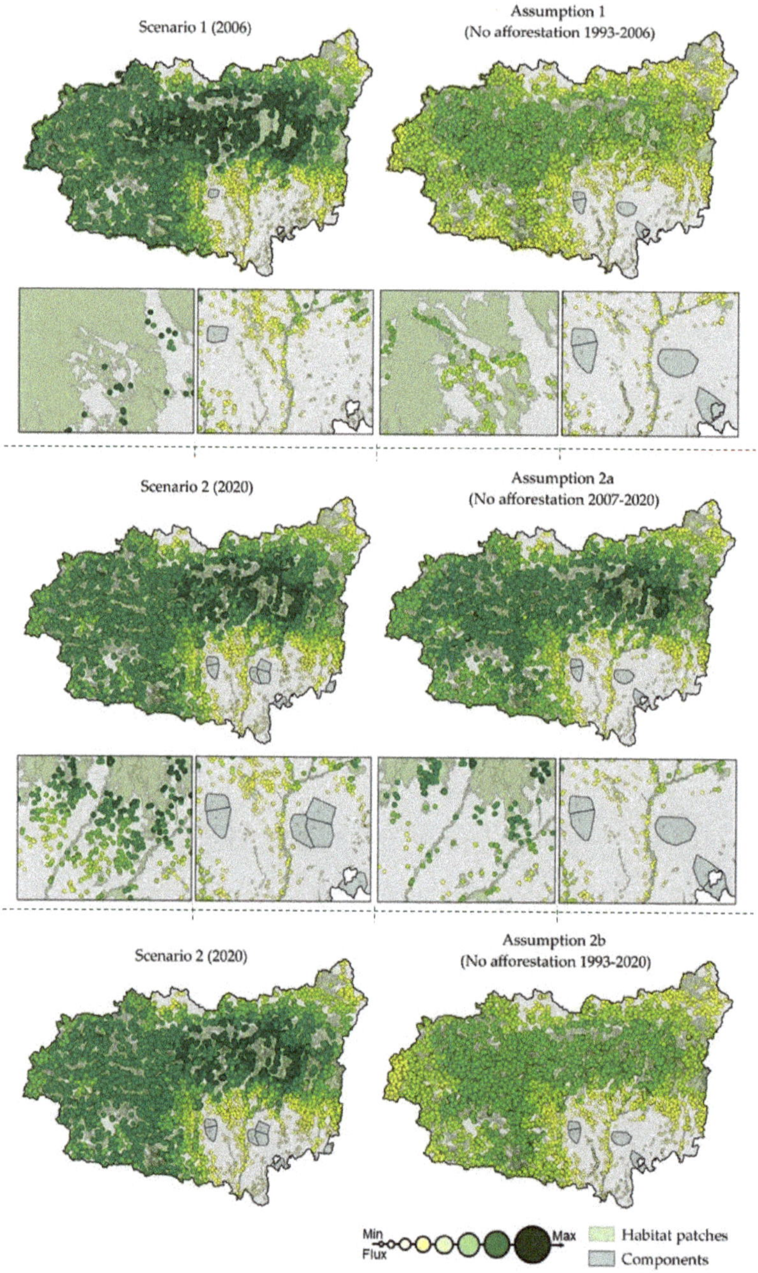

Figure 5. Evolution of forest patch connectivity in different scenarios. The size of the circle indicates the connectivity (Flux), with green shades indicating more connectivity. The native forest areas remaining in each year are shown in green.

In Assumption 2a of null afforestation from 2006 to 2020, the global connectivity metrics showed a slight decrease in the PC value (5%), global Flux (11%), and ECA (2.5%) compared to the scenario without afforestation. As for the number of components, this assumption presented two fewer components (Table 5). All isolated components, in both cases, were located in the southeastern quadrant of the study area (Figures 5 and 7). Regarding the local metrics, Flux showed a slight decrease in the central values (25–75 quartiles) and the maximum recorded value (4%), while the mean value exhibited a slight increase (2%) (Figure 6). Current Flow (CF) did not show significant differences in the distribution of patches, with an increased potential for corridors between scenario 2 and the assumption without afforestation (Figure 7). In both cases, there was a main patch, with its centroid located to the west of the central zone, highlighted by its high CF. In the assumption without afforestation, this main patch had a 20% lower CF value. This pattern occurred for the remaining patches that concentrated the potential for zonal corridors, showing lower values in the assumption without afforestation (Figure 7). The connectivity potential between patches (CF edges) was stronger in the scenario with afforestation (Figure 7).

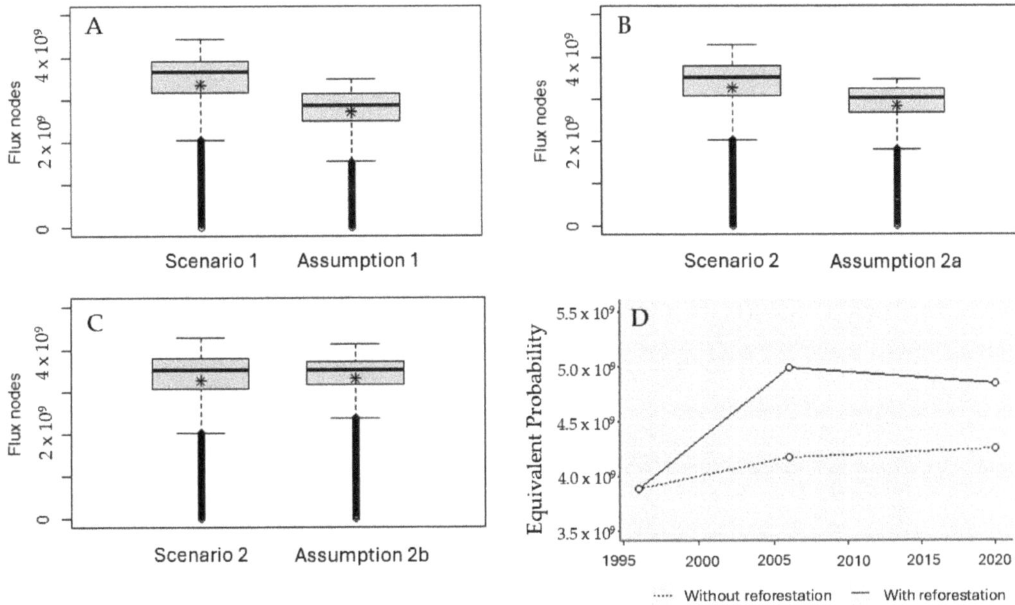

Figure 6. Effects of afforestation on connectivity metrics: (**A–C**) local Flux metric. (**D**) ECA metric. Scenario 0 (1996), scenario 1 (2006), scenario 2 (2020), Assumption 1 (no afforestation, 1993–2006), Assumption 2a (no afforestation, 2007–2020), Assumption 2b (no afforestation, 1993–2020). (*) Statistically significant difference (Student's *t*-test) at $p \leq 0.05$.

The analysis of the number of links and their costs (Table 6) showed that afforestation was conducted between 1993 and 2006 (Assumption 1: no afforestation and Assumption 2b: no afforestation) reduced fragmentation and consolidated forest patches. This led to a decrease in the number of links and an increase in travel costs as forest patches became more distant. Conversely, afforestation carried out between 2007 and 2020 (Assumption 2a) was more dispersed, which further reduced the number of links.

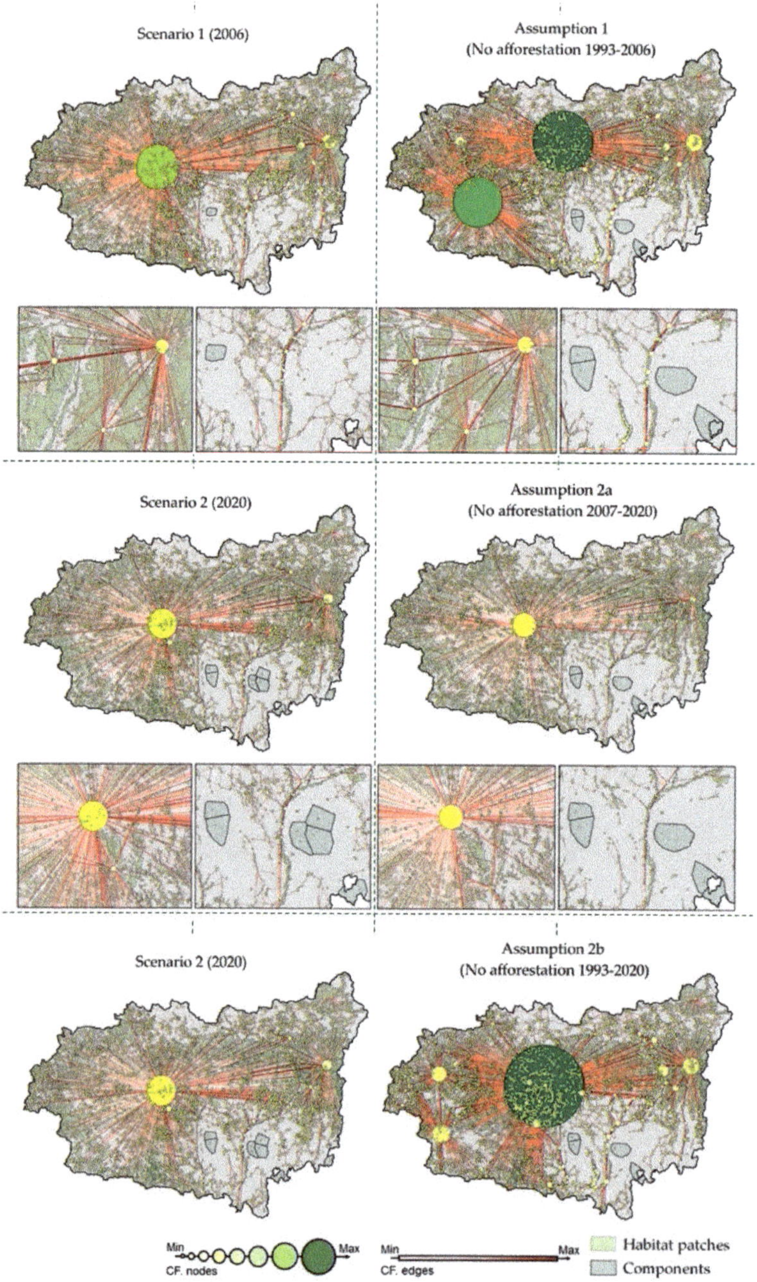

Figure 7. Evolution of forest patch connectivity in different scenarios. The size of the circle represents the connectivity (metric CF), with green colors indicating more connectivity. The native forest areas remaining in each year are shown in green.

Table 6. Mean and total distances in cost in and numbers of links for the different scenarios.

Scenario	Number of Links	Average Cost	Total Cost
Scenario 0 (1996)	694	441	305,892
Scenario 1 (2006)	12,562	212	2,660,843
Assumption 1 (no afforestation 1993–2006)	16,714	101	1,683,705
Scenario 2 (2020)	10,023	221	2,216,034
Assumption 2a (no afforestation 2007–2020)	8366	154	1,290,409
Assumption 2b (no afforestation 1993–2020)	17,796	95	1,696,760

4. Discussion

The landscape of the study area exhibited a dichotomy in terms of ecological connectivity, primarily influenced by topography. In mountain areas, characterized by dense forests and low human pressure, the ecological connectivity was high. In contrast, the agricultural plateau had limited connectivity due to sparse forest cover and greater human intervention. High mountain areas, with minimal representation, lacked available habitats. These results agreed with the WWF Spain report [32], which distinguished mountain massifs as ecological corridors from agricultural areas with weak connections.

4.1. Dynamics of Land Changes

Ecological connectivity directly impacts biodiversity, and small habitat patches can act as dispersal points for mobile species [15,33,34]. However, there is an ongoing debate on whether small, clustered patches are more effective than a single large patch for maintaining biodiversity. According to Fahrig (2020) [35], a cluster of small patches can support greater biodiversity than a single large patch. Additionally, forest species can adapt well to fragmented and heterogeneous agricultural landscapes [36]. The evolution of land covers in the study area followed similar trends to those observed across the Iberian Peninsula over the past four decades [37]. Although agricultural land use expanded, this expansion was moderate in the most recent period, being concentrated on areas with favorable conditions. The area dedicated to artificial uses also doubled, but this expansion had a minimal impact due to the low population density in the region. Forest systems experienced a significant expansion between 1996 and 2006, driven by rural depopulation, the abandonment of traditional activities, and afforestation subsidies [6,37]. However, this expansion did not significantly alter the overall proportion of land use due to the simultaneous growth of agricultural lands [3]. Between 2007 and 2020, forested areas, particularly mixed forests, slightly decreased, due to factors such as wildfires [38]. Forest plantations established before 2006 are still in the shrubland phase. The forest expansion observed in the late 20th century stabilized around the period from 2000 to 2006, marking the end of a long-term expansion process that began in the 20th century [6]. Continuous monitoring is required to assess potential future declines.

4.2. Fragmentation and Connectivity Analyses

Between 1996 and 2006, the study area experienced a significant increase in forest fragmentation. This result was inconsistent with the characteristics of the region, where forest patches in rural areas tend to homogenize and expand [39]. Despite the observed increase in fragmentation, the number of components decreased, suggesting that new patches contributed to the elimination of previously disconnected areas. Although fragmentation could be attributed to factors such as wildfires and transitions from afforestation to shrublands, differences in the resolution of cartographic data are likely the primary cause of the observed increase [40].

During the 2006–2020 period, fragmentation exhibited a declining trend, particularly within larger forest cores. This reduction in fragmentation correlated with habitat loss during the second period. The data suggest that smaller forest fragments were disproportionately affected, leading to a reduction in the number of patches and an increase in their average size. In these smaller patches, which may have served as stepping stones, the number of components increased four times (Table 4). Such patches are more vulnerable to edge-related disturbances [41,42] and increased livestock activity, which hinders regeneration [43]. Moreover, the matrix surrounding the patches has become increasingly impermeable due to the expansion of agricultural lands, and the decline of natural habitats such as shrublands and grassland (Figure 2) hindered connectivity.

Our results showed that global connectivity metrics, such as the Probability of Connectivity (PC) and the Equivalent Connected Area (ECA), correlated with available habitat, displaying similar trends. This reflects the mathematical importance of patch capacity, or habitat availability, in these metrics (Table 2). However, these metrics tend to underestimate connectivity when patch sizes are small [30], despite the fundamental role of small patches in maintaining connectivity [44].

Between 2007 and 2020, the connectivity metrics slightly declined, coinciding with the loss of available habitat, though to a lesser extent than in the first period. An increase in the global Flux metric under scenarios with reduced habitat (2b) could be explained by increased fragmentation, which incremented the number of links between patches while reducing the average and cumulative dispersal costs (Table 6). The global Flux metric was influenced by dispersal potential (Table 2), which may have facilitated the colonization of new habitats and species dispersal, potentially reducing the permeability of the study area.

These findings suggest that, as previously reported, improvements in ecological connectivity have stagnated in the early decades of the 21st century, potentially due to the transition from forest plantations to shrublands [6]. However, further studies are needed to confirm this trend.

4.3. Contribution of Afforestation Programs and Conservation Implications

Between 1993 and 2020, afforestation efforts significantly contributed to increased ecological connectivity, as evidenced by both global metrics (Table 3) and local metrics (Figures 5–7). However, the magnitude of this increase varied with the metric analyzed. Afforestation projects implemented during the first period (1993–2006) had a significant impact on the cohesion of forest masses, reduced the number of components, increased average patch size, and decreased the number of patches, particularly in mountain areas with forest potential [45]. These results are in line with the objectives of these afforestation initiatives, which aimed to homogenize and expand forested areas to facilitate management.

In contrast, afforestation efforts during the second period (2007–2020) were less substantial, more distant from existing patches (increasing the number of components), and primarily targeted agricultural areas. These results increased landscape heterogeneity, expanded connections, and enhanced dispersal potential (Figures 4 and 6). Many afforestation projects were implemented in areas distant from pre-existing forest masses, as evidenced by the increased average dispersal cost (Table 6). Additionally, the rise in the number of components suggests that new forest plantations were established in previously disconnected areas, particularly agricultural zones. While these new patches contribute little to global connectivity, they can connect isolated areas, serving as stepping stones between different habitats, as reflected in our CF results.

Afforestation programs in agricultural areas, particularly those supported by the European Agricultural Fund for Rural Development, played a crucial role in enhancing forest connectivity. The initial programs had the greatest impact (Figure 5 and

Tables 2 and 3) [3,46]. Despite improvements in functional connectivity in mountain areas, further analysis is needed to understand the impact of those programs on biodiversity and landscape configuration. Some authors have suggested that afforestation efforts on abandoned lands in various regeneration stages may not have promoted the ecological heterogeneity required for greater biodiversity [46]. Therefore, while afforestation contributes to ecological connectivity, its role in agricultural landscapes could be more important than in forest landscapes for conservation and restoration, mitigating the polarization between agricultural and forested landscapes.

Spain presents a unique case that is particularly interesting for examining the development and implications of forest management practices, given the remarkable increase in forest cover over the past 150 years: from 12.5% of the national territory in the mid-19th century to over 50% today [47], with the consequent increase in habitat availability for forest species.

Fragmentation and habitat loss are correlated with significant declines in biodiversity [48,49]. Habitat reduction decreases the persistence of animals within fragments, while increased isolation limits movement between fragments, thereby reducing recolonization following local extinctions. Decreased fragment size and increased isolation generally result in lower abundances of birds, mammals, insects, and plants [50,51]. Fragmentation impacts species differently, depending on mobility, habitat specificity, and other traits. Larger-bodied species, terrestrial and arboreal mammals, and forest-dependent species are more negatively affected by fragmentation [52,53].

Thus, increased forest area, reduced fragmentation, and improved connectivity are expected to benefit forest species. This trend was evident in our study area. For example, the Iberian brown bear (*Ursus arctos pyrenaicus*), classified as "Endangered" in 2009, was historically confined to two isolated strongholds: the Pyrenees and the Cantabrian Mountain Range (where our study area is located). Moreover, the Cantabrian population was divided into two genetically disconnected subpopulations [54]. Recent reports have indicated population growth, accompanied by an increase in migrant males, reflecting genetic connectivity between populations [55]. These results suggest that dispersed forest plantations are effectively enhancing population connectivity, which is in line with our connectivity results [33].

However, this reduction in fragmentation has still not benefited species requiring less fragmented habitats. The Cantabrian capercaillie (*Tetrao urogallus cantabricus*), a critically endangered species with only about two hundred individuals left, is severely threatened by fragmentation [56]. Habitat fragmentation forces this species into open areas, where it is more exposed to predation. The species requires approximately 500 hectares of forest habitat, but many current forest fragments do not exceed 200 hectares [57,58].

Although this approach offers several advantages, it also has some limitations, primarily due to the limited availability of data for cost evaluation. To address this issue, our study assigned generic costs to generic species [15,59]. Additionally, temporal comparisons of the layers are challenging due to differences in pixel size; however, this issue was mitigated by standardizing pixel resolution. Another limitation was the lack of data on afforestation success since all reforested areas are assumed to be successful. Nevertheless, evaluating afforestation success was beyond the scope of this study, which focused on assessing afforestation effects on fragmentation and connectivity.

5. Conclusions

In the first period studied, the forest area in the province of León increased under the influence of the first two afforestation programs. However, its evolution reached a turning point in the early years of the new century. The forests of the province of León experienced a decline, despite the afforestation conducted during that period.

The afforestation programs studied play a significant role in the evolution of the state of connectivity of forest masses. Afforestation in mountain areas contributed to the continuity of tree cover, resulting in a more homogeneous and less diverse landscape, while in the agrarian land, it generated new components and diversified the landscape.

The set of afforestation programs have had a significant impact on the state of connectivity of forest masses, landscape configuration, and diversity. The legislation of the afforestation programs, supported by technical and ecological criteria, could be a key tool to face the immediate challenges in land management.

Author Contributions: R.M.N.-C., J.B.N. and C.A.R. conceived and designed the experiments. J.B.N., C.A.R. and R.M.N.-C. performed the experiments. J.B.N., C.A.R. and R.M.N.-C. analyzed the data. J.B.N., G.P.-R., C.A.R. and R.M.N.-C. wrote the manuscript; all the authors provided editorial advice. All authors have read and agreed to the published version of the manuscript.

Funding: This research received no external funding.

Data Availability Statement: Data available upon request to the authors.

Conflicts of Interest: The authors declare no conflict of interest.

References

1. McKinney, M.L. Urbanization as a Major Cause of Biotic Homogenization. *Biol. Conserv.* **2006**, *127*, 247–260. [CrossRef]
2. Valiente-Banuet, A.; Verdú, M. Human Impacts on Multiple Ecological Networks Act Synergistically to Drive Ecosystem Collapse. *Front. Ecol. Environ.* **2013**, *11*, 408–413. [CrossRef]
3. Vadell, E.; de Miguel, S.; Centeno, G.F.; Robla, E.; Lerner, M.; García, J.P. La Forestación de Tierras Agrícolas: Balance de Un Instrumento de Política Forestal Para El Cambio Del Uso de La Tierra. *Cuad. Soc. Española Cienc. For.* **2019**, *45*, 1–20. [CrossRef]
4. Navarro-Cerrillo, R.M.; Rivas, C.A.; Quinto, L.; Navarro, S.H.; Varo-Martínez, M.Á.; Palacios-Rodríguez, P. Afforestation on Agricultural Land in Southern Spain: An Important Driver to Improve Forest Landscape Connectivity. *New For.* **2022**, *54*, 1061–1084. [CrossRef]
5. Ibáñez, I.; Katz, D.S.W.; Peltier, D.; Wolf, S.M.; Connor Barrie, B.T. Assessing the Integrated Effects of Landscape Fragmentation on Plants and Plant Communities: The Challenge of Multiprocess–Multiresponse Dynamics. *J. Ecol.* **2014**, *102*, 882–895. [CrossRef]
6. Saura, S.; Gonzales, S.; Elena-Roselló, R. Evaluación de Los Cambios En La Conectividad de Los Bosques: El Índice Del Área Conexa Equivalente y Su Aplicación a Los Bosques de Castilla y León. *Rev. Montes* **2011**, *106*, 15–21.
7. Mina, M.; Messier, C.; Duveneck, M.; Fortin, M.J.; Aquilué, N. Network Analysis Can Guide Resilience-Based Management in Forest Landscapes under Global Change. *Ecol. Appl.* **2021**, *31*, e2221. [CrossRef]
8. Kool, J.T.; Moilanen, A.; Treml, E.A. Population Connectivity: Recent Advances and New Perspectives. *Landsc. Ecol.* **2013**, *28*, 165–185. [CrossRef]
9. United Nations Environment Programme (UNEP) Unep Launches Global Connectivity Conservation Project—UNEP-WCMC. Available online: https://www.unep-wcmc.org/news/unep-launches-global-connectivity-conservation-project (accessed on 17 June 2021).
10. Dickson, B.G.; Albano, C.M.; Beier, P. Circuit-Theory Applications to Connectivity Science and Conservation. *Conserv. Biol.* **2019**, *33*, 239–249. [CrossRef]
11. Saura, S.; Lidón, R. A Common Currency for the Different Ways in Which Patches and Links Can Contribute to Habitat Availability and Connectivity in the Landscape. *Ecography* **2010**, *33*, 523–537. [CrossRef]
12. Yuan, R.; Zhang, N.; Zhang, Q. The Impact of Habitat Loss and Fragmentation on Biodiversity in Global Protected Areas. *Sci. Total Environ.* **2024**, *931*, 173004. [CrossRef]
13. Tischendorf, L.; Fahrig, L. How Should We Measure Landscape Connectivity? *Landsc. Ecol.* **2000**, *15*, 633–641. [CrossRef]
14. Cidre-González, A.; Rivas, C.A.; Navarro-Cerrillo, R.M. Fragmentation and Connectivity in Dehesa Ecosystems Associated with Cerambyx Spp. Dispersion and Control: A Graph-Theory Approach. *Forests* **2024**, *15*, 648. [CrossRef]
15. Rivas, C.A.; Guerrero-Casado, J.; Navarro-Cerrillo, R.M. Assessment of Habitat Connectivity in a Highly Fragmented Ecosystem: The Seasonal Tropical Dry Forest in Ecuador. *Appl. Veg. Sci.* **2024**, *27*, e12770. [CrossRef]
16. Rivas, C.A.; Guerrero-Casado, J.; Navarro-Cerrillo, R.M. Functional Connectivity across Dominant Forest Ecosystems in Ecuador: A Major Challenge for a Country with a High Deforestation Rate. *J. Nat. Conserv.* **2024**, *78*, 126549. [CrossRef]

17. Dilts, T.E.; Weisberg, P.J.; Leitner, P.; Matocq, M.D.; Inman, R.D.; Nussear, K.E.; Esque, T.C. Multiscale Connectivity and Graph Theory Highlight Critical Areas for Conservation under Climate Change. *Ecol. Appl.* **2016**, *26*, 1223–1237. [CrossRef]
18. Rivas, C.A.; Guerrero-Casado, J.; Navarro-Cerillo, R.M. Deforestation and fragmentation trends of seasonal dry tropical forest in Ecuador: Impact on conservation. *For. Ecosyst.* **2021**, *8*, 46. [CrossRef]
19. Saura, S.; Pascual-Hortal, L. A New Habitat Availability Index to Integrate Connectivity in Landscape Conservation Planning: Comparison with Existing Indices and Application to a Case Study. *Landsc. Urban Plan.* **2007**, *83*, 91–103. [CrossRef]
20. Gurrutxaga, M.; Rubio, L.; Saura, S. Key Connectors in Protected Forest Area Networks and the Impact of Highways: A Transnational Case Study from the Cantabrian Range to the Western Alps (SW Europe). *Landsc. Urban Plan.* **2011**, *101*, 310–320. [CrossRef]
21. Beier, P.; Majka, D.R.; Spencer, W.D. Forks in the Road: Choices in Procedures for Designing Wildland LinkagesBifurcaciones En El Camino: Opciones de Procedimientos Para El Diseño de Enlaces de Tierras Silvestres. *Conserv. Biol.* **2008**, *22*, 836–851. [CrossRef] [PubMed]
22. Virgós, E.; Tellería, J.L.; Santos, T. A Comparison on the Response to Forest Fragmentation by Medium-Sized Iberian Carnivores in Central Spain. *Biodivers. Conserv.* **2002**, *11*, 1063–1079. [CrossRef]
23. Virgós, E.; García, F.J. Patch Occupancy by Stone Martens Martes Foina in Fragmented Landscapes of Central Spain: The Role of Fragment Size, Isolation and Habitat Structure. *Acta Oecologica* **2002**, *23*, 231–237. [CrossRef]
24. Virgós, E. Tejón—Meles Meles. In *Enciclopedia Virtual de los Vertebrados Españoles*; Salvador, A., Cassinello, J., Eds.; Museo Nacional de Ciencias Naturales: Madrid, Spian, 2012.
25. Foltête, J.C.; Clauzel, C.; Vuidel, G. A Software Tool Dedicated to the Modelling of Landscape Networks. *Environ. Model. Softw.* **2012**, *38*, 316–327. [CrossRef]
26. Foltête, J.C.; Clauzel, C.; Vuidel, G.; Tournant, P. Integrating Graph-Based Connectivity Metrics into Species Distribution Models. *Landsc. Ecol.* **2012**, *27*, 557–569. [CrossRef]
27. Clauzel, C.; Foltête, J.-C.; Girardet, X.; Vuidel, G. *User Manual Graphab 2.4*; University of Franche-Comté—CNRS: Besançon, France, 2019.
28. Urban, D.; Keitt, T. Landscape Connectivity: A Graph-Theoretic Perspective. *Ecology* **2001**, *82*, 1205–1218. [CrossRef]
29. Saura, S.; Torné, J. Conefor Sensinode 2.2: A Software Package for Quantifying the Importance of Habitat Patches for Landscape Connectivity. *Environ. Model. Softw.* **2009**, *24*, 135–139. [CrossRef]
30. Saura, S.; Estreguil, C.; Mouton, C.; Rodríguez-Freire, M. Network analysis to assess landscape connectivity trends: Application to European forests (1990–2000). *Ecol. Indic.* **2010**, *11*, 407–416. [CrossRef]
31. Girardet, X.; Foltête, J.C.; Clauzel, C. Designing a Graph-Based Approach to Landscape Ecological Assessment of Linear Infrastructures. *Environ. Impact Assess. Rev.* **2013**, *42*, 10–17. [CrossRef]
32. Rodríguez, G.; Mateo Sánchez, M.C.; de la Fuente Martín, B.; Gastón González Santiago, A.; Saura Martínez, S.; Gurrutxaga, M. *Autopistas Salvajes. Propuesta de WWF España Para Una Red Estratégica de Corredores Ecológicos Entre Espacios Red Natura 2000*; WWF/Adena: Madrid, Spain, 2018.
33. Diniz, M.F.; Coelho, M.T.P.; de Sousa, F.G.; Hasui, É.; Loyola, R. The Underestimated Role of Small Fragments for Carnivore Dispersal in the Atlantic Forest. *Perspect. Ecol. Conserv.* **2021**, *19*, 81–89. [CrossRef]
34. Herrera, L.P.; Sabatino, M.C.; Jaimes, F.R.; Saura, S. Landscape Connectivity and the Role of Small Habitat Patches as Stepping Stones: An Assessment of the Grassland Biome in South America. *Biodivers. Conserv.* **2017**, *26*, 3465–3479. [CrossRef]
35. Fahrig, L. Why Do Several Small Patches Hold More Species than Few Large Patches? *Glob. Ecol. Biogeogr.* **2020**, *29*, 615–628. [CrossRef]
36. Gastón, A.; Blázquez-Cabrera, S.; Garrote, G.; Mateo-Sánchez, M.C.; Beier, P.; Simón, M.A.; Saura, S. Response to Agriculture by a Woodland Species Depends on Cover Type and Behavioural State: Insights from Resident and Dispersing Iberian Lynx. *J. Appl. Ecol.* **2016**, *53*, 814–824. [CrossRef]
37. Fernández Nogueira, D.; Corbelle Rico, E. Cambios En Los Usos de Suelo En La Península Ibérica: Un Meta-Análisis Para El Período 1985–2015. *Rev. Bibliográfica Geogr. Cienc. Soc.* **2017**, *XXII*.
38. Molinero Hernando, F.; Guerra Velasco, J.C.; de Cascos Maraña, C.S. La Dinámica de Los Incendios Forestales En Castilla y León Como Resultado Del Abandono y La Despoblación Durante El Último Cuarto de Siglo. In *Investigando en Rural*; Coloquio de Geografía Rural; Ulzama Ediciones: Ulzama, Spain, 2012; pp. 473–482, ISBN 978-84-92870-81-3.
39. Regato, P.; Castejón, M.; Tella, G.; Giménez, S.; Barrera, I.; Elena-Rosselló, R. Cambios Recientes En Los Paisajes de Los Sistemas Forestales Mediterráneos de España. *Inv. Agr. Sist. Rec. For.* **1999**, *1*, 383–398.
40. Hengl, T. Finding the Right Pixel Size. *Comput. Geosci.* **2006**, *32*, 1283–1298. [CrossRef]
41. Banks-Leite, C.; Ewers, R.M.; Folkard-Tapp, H.; Fraser, A. Countering the Effects of Habitat Loss, Fragmentation, and Degradation through Habitat Restoration. *One Earth* **2020**, *3*, 672–676. [CrossRef]
42. Franklin, C.M.A.; Filicetti, A.T.; Nielsen, S.E. Seismic Line Width and Orientation Influence Microclimatic Forest Edge Gradients and Tree Regeneration. *For. Ecol. Manag.* **2021**, *492*, 119216. [CrossRef]

43. Pedley, D.; McWilliam, W.; Doscher, C. Forests from the Grass: Natural Regeneration of Woody Vegetation in Temperate Marginal Hill Farmland under Minimum Interference Management. *Restor. Ecol.* **2023**, *31*, e13852. [CrossRef]
44. Ramírez-Delgado, J.P.; Di Marco, M.; Watson, J.E.M.; Johnson, C.J.; Rondinini, C.; Corredor Llano, X.; Arias, M.; Venter, O. Matrix Condition Mediates the Effects of Habitat Fragmentation on Species Extinction Risk. *Nat. Commun.* **2022**, *13*, 595. [CrossRef]
45. Vadell, E.; Pemán, J.; Verkerk, P.J.; Erdozain, M.; de-Miguel, S. Forest Management Practices in Spain: Understanding Past Trends to Better Face Future Challenges. *For. Ecol. Manag.* **2022**, *524*, 120526. [CrossRef]
46. García, A.D.d.C.; Alcázar, J.M.G.; Cerrillo, R.M.N. Análisis Retrospectivo de Las Reforestaciones En Tierras Agrarias En El Municipio de Tembleque (Toledo). *Cuad. Soc. Española Cienc. For.* **2008**, *28*, 145–150.
47. de Armenteras, A.A. *Árboles y Montes: Curiosidades Artísticas é Históricas de Los Montes: Con La Explicación de Las Más Beneficiosas Influencias Del Arbolado y de Las Más Importantes Nociones Forestales*; Ricardo Rojas: Madrid, Spain, 1903.
48. Rogan, J.E.; Lacher, T.E. *Impacts of Habitat Loss and Fragmentation on Terrestrial Biodiversity*; Reference Module in Earth Systems and Environmental Sciences; Elsevier: Amsterdam, The Netherlands, 2018. [CrossRef]
49. Betts, M.G.; Wolf, C.; Pfeifer, M.; Banks-Leite, C.; Arroyo-Rodríguez, V.; Ribeiro, D.B.; Barlow, J.; Eigenbrod, F.; Faria, D.; Fletcher, R.J.; et al. Extinction Filters Mediate the Global Effects of Habitat Fragmentation on Animals. *Science* **2019**, *366*, 1236–1239. [CrossRef]
50. Haddad, N.M.; Gonzalez, A.; Brudvig, L.A.; Burt, M.A.; Levey, D.J.; 1Department, E.I.D. Experimental Evidence Does Not Support the Habitat Amount Hypothesis. *Ecography* **2017**, *40*, 48–55. [CrossRef]
51. Haddad, N.M.; Brudvig, L.A.; Clobert, J.; Davies, K.F.; Gonzalez, A.; Holt, R.D.; Lovejoy, T.E.; Sexton, J.O.; Austin, M.P.; Collins, C.D.; et al. Habitat Fragmentation and Its Lasting Impact on Earth's Ecosystems. *Sci. Adv.* **2015**, *1*, e1500052. [CrossRef]
52. Lino, A.; Fonseca, C.; Rojas, D.; Fischer, E.; Ramos Pereira, M.J. A Meta-Analysis of the Effects of Habitat Loss and Fragmentation on Genetic Diversity in Mammals. *Mamm. Biol.* **2019**, *94*, 69–76. [CrossRef]
53. Amos, J.N.; Harrisson, K.A.; Radford, J.Q.; White, M.; Newell, G.; Nally, R.M.; Sunnucks, P.; Pavlova, A. Species- and Sex-Specific Connectivity Effects of Habitat Fragmentation in a Suite of Woodland Birds. *Ecology* **2014**, *95*, 1556–1568. [CrossRef] [PubMed]
54. Grupo de Trabajo del Oso Pardo Pirenaico. In *Estrategia Para la Conservación del Pardo (Ursus Arctos) en los Pirineos*; Ministerio de Medio Ambiente y Medio Rural y Marino, Secretaría General Técnica, Centro de Publicaciones: Madrid, Spain, 2006.
55. Fundación Oso Pardo Nuevos Tiempos Para Las Poblaciones Oseras Ibéricas. *Ambient. Rev. Minist. Medio Ambiente* **2017**, *118*, 50–57.
56. García-Rodríguez, A.; Herrero-García, G.; de Garnica García, M.G.; García Esgueva, Á.; Balsera, R.; Oleaga, Á.; Fernández, D.; Amado, J.; Royo, L.J.; García Iglesias, M.J.; et al. Mortality Causes in Captive Cantabrian Capercaillie (Tetrao Urogallus Cantabricus) in Spain. *Animals* **2023**, *13*, 1255. [CrossRef] [PubMed]
57. SEO Birdlife. Urogallo Común. Available online: https://seo.org/ave/urogallo-comun/ (accessed on 12 December 2024).
58. SEO BirLife. *Libro Rojo de Las Aves de España*; Madroño, A., Gonzáles, C., Atienza, J.C., Eds.; Miniserio de Medio Ambiente: Madrid, Spina, 2004.
59. Hilty, J.; Worboys, G.L.; Keeley, A.; Woodley, S.; Lausche, B.J.; Locke, H.; Carr, M.; Pulsford, I.; Pittock, J.; White, J.W.; et al. *Guidelines for Conserving Connectivity Through Ecological Networks and Corridors*; IUCN: Gland, Switzerland, 2020.

land

MDPI

Article

How Are Residents' Livelihoods Affected by National Parks? A SEM Model Based on DFID Framework

Likun Gu [1,2,3,*], Guoqing Shi [1,2,3,*], Yuanke Zhao [1,2,3], Huicong Liu [1,2,3] and Xinyu Ye [1,2,3]

[1] National Research Center for Resettlement, Hohai University, Nanjing 211100, China; zyk@hhu.edu.cn (Y.Z.); huicongliu@hhu.edu.cn (H.L.); 231614010059@hhu.edu.cn (X.Y.)
[2] Institute of Social Development, Hohai University, Nanjing 211100, China
[3] Asian Research Center, Hohai University, Nanjing 211100, China
* Correspondence: 200414120001@hhu.edu.cn (L.G.); gshi@hhu.edu.cn (G.S.)

Abstract: National parks represent a global initiative for biodiversity conservation and environmentally sustainable societal development, with China having launched its own national park program. The establishment and operation of these parks significantly impact local residents' livelihoods. Based on DFID's Sustainable Livelihoods Framework, an assessment tool introduced by the UK Department for International Development (DFID) for evaluating the livelihood standards of residents, this study constructs a structural equation modeling (SEM) framework to analyze how national parks affect residents' livelihoods, discussing livelihood risk management and feasible capacity-building interventions. Focusing on the Northeast Tiger and Leopard National Park as a case study, the research reveals that indirect wildlife-inflicted damage poses more pronounced negative impacts on local communities than park establishment policies. Both regulatory land-use restrictions and wildlife conflicts disrupt land-based livelihood activities, ultimately affecting residents' livelihood stability. Mitigation requires comprehensive measures, including retaining essential farmland; providing vocational skill training; offering specialized loans; diversifying employment channels; and improving compensation mechanisms to safeguard residents' livelihood security.

Keywords: national park; sustainable livelihood; SEM model; livelihood framework; human–animal conflicts

1. Introduction

National parks refer to specific terrestrial or marine areas with clearly defined boundaries approved and centrally managed by the state whose primary purpose is to protect large-scale natural ecosystems of national significance [1]. They aim to achieve scientific conservation and the rational utilization of natural resources. As a type of global protected area, the fundamental attributes of national parks lie in their public welfare nature, state leadership, and scientific management [2]. As a proven and effective model for natural ecosystem conservation, national parks have now been widely adopted in the majority of countries and regions worldwide [3–5].

However, it is undeniable that, while the establishment of national parks represents a quintessential environmentally beneficial initiative, it inevitably entails certain ecological and social challenges [6]. Among these, the most prominent issue is residents' livelihoods [7]. This arises because national parks are typically established subsequent to existing human activities and communities, generally triggering competing conflicts between the national park and local residents over the fundamental resource of land. Current

 https://doi.org/10.3390/land14071501

research on this topic primarily focuses on national parks in regions such as Africa and Southeast Asia. The Awash National Park in Ethiopia [3], the Doi Inthanon National Park in Thailand [4], the Mount Elgon National Park in Uganda [5], and other national parks in different areas have attracted widespread attention on this topic. This concentration stems primarily from the underdeveloped socioeconomic conditions in these areas, resulting in more pronounced impacts of national parks on local communities' livelihoods and, consequently, generating greater social risks [5,8].

At the inception of Yellowstone National Park—the world's first national park—its establishment aimed solely at preventing human disturbance to protected wilderness [9]. In the 20th century, the exclusionary nature of national parks toward human activities garnered widespread attention. Park founders and administrators gradually recognized the problems arising from such exclusion, including resettlement conflicts [10], community disintegration, and livelihood deterioration [11,12]. Ironically, the emergence and escalation of these issues threatened the integrity of national parks themselves, while diminishing environmental benefits due to the resultant social problems. Consequently, numerous governments began safeguarding the rights of residents within national park boundaries during park establishment [13]. Driven by the diffusion and evolution of this humanistic philosophy, balancing environmental and social benefits became a global consensus in national park development [14]. Particularly noteworthy is the geographic overlap between proposed national park areas and regions characterized by traditional ethnic cultures or widespread livelihood poverty. This convergence necessitates that national park development advance not only ecological sustainability but also social sustainability [7].

When addressing this issue, protecting the livelihoods of resident indigenous communities at the community level becomes a crucial consideration in the establishment of ecological conservation areas represented by national parks [15]. Such livelihood protection must account for both preserving the current standard of living and enhancing future livelihood opportunities [16], thereby achieving sustainability over time. The assessment of livelihood sustainability does not hinge on whether relocation occurs but is measured by the degree of impairment to livelihood capitals, which determines the resilience of indigenous residents' livelihoods [17].

Compared to other nation-states in the international community, China embarked on its national park development initiative at a comparatively later historical juncture [18]. Following international best practices, China has begun integrating previously fragmented and functionally limited protected areas through its national park initiative [19]. However, as a developing nation with a vast territory and extraordinary biodiversity, China faces heightened challenges in reconciling ecological conservation goals with local socioeconomic development. Furthermore, it is impractical to achieve the wholesale relocation of resident communities within designated protected areas due to competing considerations of investment requirements, economic repercussions, social consequences and risks, ethnocultural preservation needs, and natural resource utilization imperatives [20,21]. Consequently, achieving balanced sustainable development between national parks and resident livelihoods remains a critical challenge demanding urgent research [22]. These recurring scenarios across populous developing countries signify that China's national park development approaches and insights will constitute a particularly valuable component of global conservation knowledge systems.

In 2017, China issued the "Master Plan for Establishing the National Park System", then quickly established the first batch of five national parks: Sanjiangyuan, Giant Panda, Northeast Tiger and Leopard, Hainan Tropical Rainforest, and Mount Wuyi in 2021 [23]. In the development of China's national parks, a similar logic existed as in other countries, positing mutual exclusivity between park establishment and the productive/livelihood activities of resident communities [24]. This is explicitly codified in the zoning provisions of the Interim Measures for National Park Management, which stipulate that core protection zones shall prohibit all human activities in principle, while general protection zones ban development-oriented production activities [25]. It is evident that such restrictive measures have imposed significant external social constraints on the local residents' productive activities and daily lives, probably resulting in negative impacts. Furthermore, one of the stated objectives of national park construction—wildlife conservation—has paradoxically led to an increased frequency of human–wildlife conflicts. Consequently, the establishment of national parks not only affects the development and utilization of land and other resources but also perpetuates risks to personal safety and property security [26]. However, it is noteworthy that China has substantially drawn lessons from past international experiences during its national park development process. All five current national park master plans in China explicitly emphasize the imperative to safeguard local residents' basic livelihood standards [27]. Through initiatives such as promoting tourism operations and forest-based agriculture, these plans aim to facilitate livelihood transformation for local communities.

Current discourse and research predominantly focus on analyzing the composition of residents' livelihood capitals while rarely systematically investigating the impact pathways through which national parks influence these capital assets. This study aims to address the following questions: (1) Which elements within the livelihood capital system of residents in national parks determine the final livelihood outcomes? (2) Between the restrictive measures imposed by the establishment of national parks and the rise in wildlife–human conflicts caused by these protected areas, which one has a more profound impact on local residents' livelihoods? (3) What are the specific pathways through which national parks influence the livelihood capital system of residents? To address these questions, this research selects the Northeast Tiger and Leopard National Park as a case study to analyze the specific mechanisms by which national parks affect the livelihood capital system of residents. Adopting a quantitative research paradigm, the study combines multiple variables to explore the driving forces behind the livelihood transformation of national parks residents. The primary contributions of this study include the following: (1) Integrating the structure of residents' livelihood capitals with risk shocks under national park influences to establish an analytical framework for livelihood risk shocks; (2) Employing structural equation modeling (SEM) to quantitatively map the causal pathways through which national park development affects residents' livelihoods; (3) Comparing the differential impact of national park establishment across various types of livelihood capital. Building on these findings could provide clarity to discuss evidence-based policies to foster the synergistic sustainability of both national parks and community livelihoods.

2. Review

2.1. Purpose and Policies of Chinese National Parks

Unlike nature reserves, which aim to protect single or a few species of wildlife or natural landscapes, national parks are characterized by their systematic protection and management of all wildlife, flora, and natural landscapes (sometimes also including cultural asset) within a designated geographical area. Therefore, the establishment of national parks signifies that China's ecological and environmental conservation efforts have officially entered a new historical phase [28].

In 2013, the Chinese government first proposed the establishment of national parks. After years of practical exploration and experience accumulation, China officially inaugurated its first batch of national parks in 2021, including Sanjiangyuan, Giant Panda, Northeast China Tiger and Leopard, Hainan Tropical Rainforest, and Wuyishan [28]. These parks span a total protected area of 230,000 km^2, covering nearly 30% of China's nationally protected terrestrial wildlife species. Through the creation and refinement of the national park system, China has strengthened the conservation of rare and endangered wildlife and their habitats and ecosystems [29]. Numerous critically endangered species have achieved population recovery, with the national protection rate for key wildlife rising to 74% and biodiversity becoming significantly richer. According to the National Park Spatial Layout Plan, China aims to establish 49 national parks by 2035, forming the world's largest national park system with a total area of approximately 1.1 million km^2 [30].

While ecological and environmental conservation efforts continue to advance, it is crucial to recognize that the establishment of a national park fundamentally relies on restrictions and limitations on land use and the restriction of human activities [1,31]. Under current regulations, China's national parks are divided into core conservation zones and general control zones based on their functional priorities [22]. These two categories differ in the intensity of restrictions imposed on human activities, with the general control zones typically having a lesser impact on local residents' livelihoods and daily activities [19]. Nevertheless, due to variations in ecological functions and geographical contexts, the proportion of core conservation zones to general control zones differs across national parks. This discrepancy further manifests in the uneven demographic impact of different parks, as the number of residents affected by these zoning policies varies significantly depending on the park's specific ecological mandates and regional characteristics.

Regulatory measures in core conservation zones and general control zones vary across different national parks. For instance, Northeast Tiger and Leopard National Park, Sanjiangyuan National Park, and Giant Panda National Park primarily emphasize "no expansion of existing productive and commercial activities" within these zones. Hainan Tropical Rainforest National Park and Wuyishan National Park, however, adopt stricter policies that explicitly require the phasing out of human activities, reflecting a more stringent approach. In general control zones, current policies largely acknowledge the impracticality of the immediate and complete cessation of local livelihoods [29,32]. Consequently, these areas are granted greater policy flexibility and operational leeway. A notable exception is Hainan Tropical Rainforest National Park, where the migratory and sedentary behaviors of its flagship protected species, the Hainan gibbon (*Nomascus hainanus*), have necessitated a dynamic management policy for its general control zones. This approach adapts to the primates' ecological needs while balancing human activity regulation.

The five existing Chinese national parks currently adopt largely consistent management strategies for livelihood restoration and alternative livelihood development, primarily encouraging local communities to establish green, specialty businesses centered on unique local resources, particularly eco-tourism, leveraging ecological landscapes and cultural heritage [33]. Some parks have implemented innovative practices such as purchasing services from residents to employ them as forest rangers, which has already yielded measurable success [34]. Additionally, notable progress is being made in two key areas of ecological compensation:

(1) Wildlife Conflict Compensation System: National parks are refining mechanisms to provide financial compensation for residents' unexpected losses caused by wildlife, ensuring agricultural productivity and community engagement remain intact [35].
(2) Carbon Credit Market Participation: China's national parks are exploring entry into carbon markets, capitalizing on their ecological assets to generate self-sustaining revenue. This initiative aims to reduce fiscal reliance on central and local governments while advancing climate goals [35].

Both approaches represent critical ecological compensation mechanisms, balancing conservation objectives with socio-economic sustainability.

2.2. Sustainable Livelihood Framework

Considering that national parks, as an external factor, have an impact on the livelihood base of residents, it is necessary to assess the livelihood status of residents within national park development areas. Following the widely adopted DFID Sustainable Livelihoods Framework, the livelihood capital system of residents can be categorized by elemental characteristics and logical relationships into human capital, social capital, natural capital, physical capital, and financial capital [36,37]. This allows the determination of each livelihood capital dimension through measurable variable combinations, thereby calculating the comprehensive level of livelihood capital and evaluating residents' sustainable livelihood capacities [38]. Similarly, this framework can visualize the impacts of national park development on local residents' livelihoods.

The DFID Sustainable Livelihoods Framework (SLF) is an assessment tool introduced by the UK Department for International Development (DFID) for evaluating the livelihood standards of residents [36]. As an analytical tool, it has been extensively employed to assess livelihood capabilities [39]. Its central tenet lies in revealing the structural characteristics formed through the differentiation of livelihood assets, thereby determining communities' capacity to sustain production and daily life [40,41]. As an analytical instrument, the DFID framework predominantly focuses on statically presenting the livelihood assets portfolio [42]. By analyzing historical and current asset configurations while incorporating external risk shocks or enhancement opportunities [43], it evaluates potential impacts on future livelihood capacities and living standards [44].

The DFID Sustainable Livelihoods Framework (SLF) has been ubiquitously applied in current research to analyze livelihood disruptions caused by external natural or societal factors. Natural drivers predominantly encompass catastrophic events—such as earthquakes, typhoons, tsunamis, and floods—that induce the systemic degradation of livelihood assets. Societal drivers primarily involve large-scale infrastructure projects (e.g., reservoirs, nuclear power plants, industrial parks) leading to resettlement and asset depreciation [38,45–47]. The SLF demonstrates analytical clarity in delineating resilient versus vulnerable assets within livelihood portfolios, enabling the identification of which asset categories face heightened exposure to exogenous risks [48]. By integrating these two diagnostic procedures, researchers can map the causal chain of "External Stressors → Asset Erosion → Livelihood Capacity Deterioration", ultimately formulating targeted policy interventions and community support mechanisms. Currently, there remains a lack of such research in the context of China's national parks, with most existing studies confined to discussions on the structure of residents' livelihood capital. As a hybrid external stressor, national parks exert dual impacts: ecologically through biodiversity conservation mandates and socially via governance structures requiring state–society synergies—a duality rooted in their dual mission of ecological enhancement and polycentric institutional implementation [3].

2.3. Livelihood Risk Generated by a National Park

Risk is characterized by the uncertainty between production objectives and labor outcomes [49]. This uncertainty can be classified into uncertainties in production returns and cost uncertainties, depending on the aspects affected. For most farmers in developing countries, land serves as the core productive asset in household-based economic activities [50]. However, since land is frequently a prerequisite for large-scale infrastructure projects, conflicts occasionally arise between residents and the state over land use rights [51].

Due to the heterogeneity in risk sources, mechanisms, and impacts, there is currently no unified analytical framework for risk assessment [52]. Researchers typically categorize external risks affecting residents' productive and daily lives across dimensions such as personal health, environmental factors, employment opportunities, socioeconomic conditions, and infrastructure integrity [53–55]. These external risks often originate from two categories:

(1) Mega-projects (e.g., reservoirs, industrial facilities, protected areas, ports, transportation infrastructure);
(2) Unexpected incidents (e.g., floods, earthquakes, wildfires, nuclear accidents).

Building on the above context, this study categorizes risks by their specific forms into human capital risks, social capital risks, natural capital risks, physical capital risks, and financial capital risks, aligning with the Sustainable Livelihoods Framework (SLF) to ensure conceptual coherence and analytical consistency. When residents face such risks, their existing livelihood capital systems are negatively impacted [56]. To cope, they may either reduce expenditures, leading to a decline in living standards, or sell off personal assets to sustain their current lifestyle [57]. However, the latter strategy further depletes livelihood capital, ultimately trapping individuals in a cycle of poverty [58,59]. This situation establishes a foundational analytical framework: national parks, as external mega-projects, generate risks to local residents' livelihood capital due to their development objectives and operational requirements. This framework applies equally to analyzing social impacts arising from other large-scale infrastructure initiatives.

However, it is crucial to recognize that this risk categorization operates within the context of specific livelihood capital systems. Fundamentally, the ontological origin of risks resides in the national park establishment itself—a dual process that simultaneously excludes human activities while generating derivative wildlife disturbance incidents [60,61]. These cascading effects perpetuate asset depreciation across all livelihood capital domains. This reveals national parks as the singular exogenous risk generator, with perceived risks to human, social, natural, physical, and financial capitals merely representing differentiated manifestations of park-induced vulnerabilities within discrete asset categories.

2.4. Research Hypotheses

Based on the above analysis, existing research on China's national parks has produced numerous findings discussing residents' livelihood risks and potential losses. However, there is still insufficient exploration of how national parks influence various forms of livelihood capital through specific pathways, ultimately affecting livelihood outcomes. Moreover, little attention has been paid to the dual nature—both natural and social—of national parks' external impacts.

Therefore, this study attempts to formulate three hypotheses to investigate the specific pathways through which national parks affect residents' livelihoods. It also aims to assess the risk levels faced by different types of livelihood capital, thereby proposing targeted strategies for the generation, maintenance, and improvement of livelihood capital.

Variations in livelihood capital systems constitute the fundamental determinant of changes in residents' livelihood outcomes. Generally, higher robustness and structural balance within the livelihood asset portfolio correlate with elevated living standards and enhanced resilience to external shocks. While this causal relationship holds under controlled conditions devoid of exogenous disturbances, the intrinsic dynamics of capital transformation remain pivotal. Consequently, this study proposes the first hypothesis:

H1. *Each individual livelihood capital exerts a positive contributing effect on ultimate livelihood attainment.*

Given that livelihood risks fundamentally originate from exogenous natural and societal stressors, national park establishment can be posited as the direct exogenous driver of livelihood capacity transformations in this study. Functioning as a polyvalent external shock encompassing both ecological and institutional dimensions, national parks disrupt residents' productive and domestic systems. However, the magnitude of impact may exhibit capital-specific heterogeneity. Recognizing the causal mechanism whereby external perturbations operationalize through discrete capital erosion pathways to ultimately depress livelihood outcomes, we accordingly formulate the second hypothesis:

H2. *National parks exert negative impacts across all five core capital dimensions within the DFID framework.*

Finally, to contrast the differential impact pathways through which national parks affect livelihood outcomes, we introduce the third hypothesis:

H3. *National parks, as an exogenous institutional intervention, induce systemic deterioration in residents' livelihood standards.*

The specific structure is illustrated in Figure 1.

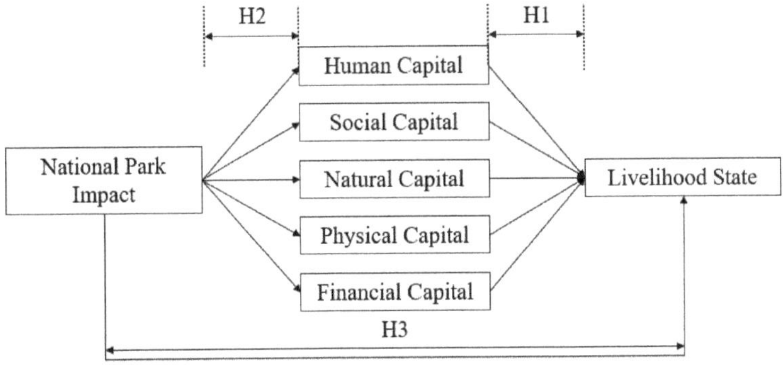

Figure 1. The structure of livelihood level.

3. Methods

3.1. Research Areas

We selected the Northeast Tiger and Leopard National Park as the study area for the following reasons: (1) This national park is located in the boundary between Heilongjiang Province and Jilin Province, which means the development of industries in different towns is different, and the laws operated by local governments would also be different. (2) The overall livelihood level of the farming community living in this area is relatively poor, and the local government, due to financial constraints, cannot provide sufficient funds to help them recover their livelihoods. (3) The wildlife protected in this national park is highly prone to causing accidents and inflicting harm, and this uncontrollable risk will continue to increase as their population grows.

The above situation indicates that the recovery of the livelihood level of local farmers faces significant risks. Therefore, we selected Northeast Tiger and Leopard National Park as a case study to analyze the internal driving mechanisms of the livelihood transformation of local farmers.

The Northeast Tiger and Leopard National Park covers a total area of 14,100 km^2 [1]. Its eastern border is adjacent to Russia's Land of the Leopard National Par", and its southeastern part faces North Korea across the Tumen River, with a border line of 256 km [62]. The administrative regions involved include 18 townships in three counties (cities) of Hunchun, Wangqing, and Tumen in Jilin Province, and 10 townships in three counties (cities) of Dongning, Muleng, and Ning'an in Heilongjiang Province (Figure 2).

Among them, the area in Jilin Province is 9557 km^2, accounting for 68% of the total area of the Northeast Tiger and Leopard National Park, while the area in Heilongjiang Province is 4508 km^2, accounting for 32% [63].

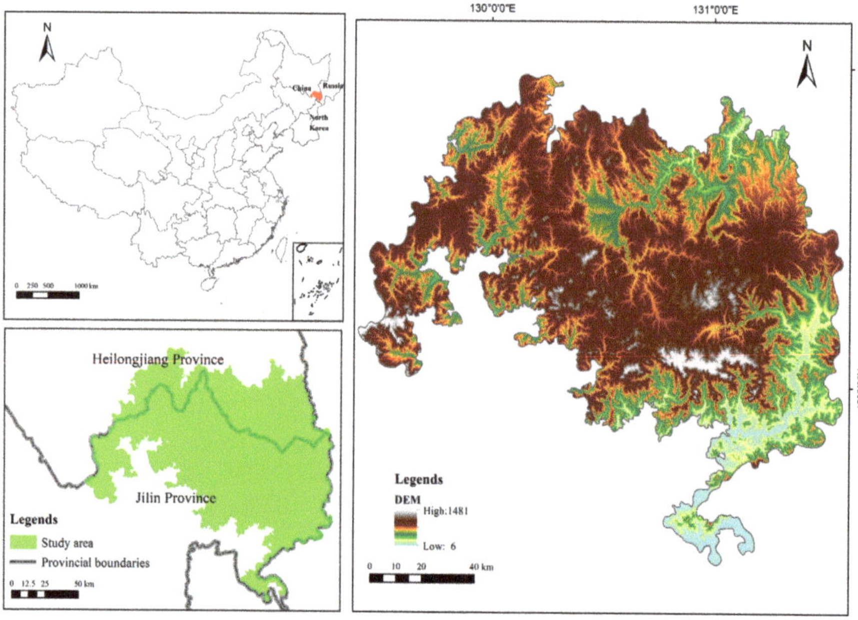

Figure 2. The map of the research area (Northeast Tiger and Leopard National Park).

In the Northeast Tiger and Leopard National Park area, there are currently 8540 registered households with a population of 19,100 people. Among them, 691 households and 15,200 people are registered under forestry management, while 1629 households and 3900 people are registered in rural villages. The population is predominantly composed of Han and Korean ethnic groups. In addition, nearly 70,000 people live outside the park, but their arable land, forest land, and other means of production are located within the park [64].

The residential areas within the park are characterized by large concentrations and small dispersals, with 12 natural villages and 43 forestry station headquarters. The total land area within the park is 14,065 km^2, of which 12,893 km^2 are state-owned, accounting for 91.7%, and 1172 km^2 are collectively owned, accounting for 8.3% [65]. The predominant land type is forest land, which makes up 97.6% of the total area, with a small amount of arable land, orchard land, water bodies, and construction land also present [66].

3.2. Data Source

In June 2024, this study distributed 642 questionnaires, with 614 returned and 567 validated. The research focused on rural community residents within the Northeast Tiger and Leopard National Park area, defining them as residents caused by national park development. The study aimed to measure the structure and intensity of livelihood capital among these residents. Data were collected using Likert scales and scoring methods to quantify livelihood status and alternative livelihood strategies. Following the Sustainable Livelihoods Framework, the livelihood levels of residents were categorized into five dimensions: human capital, social capital, natural capital, physical capital, and financial capital. The underlying logic of this analytical framework lies in treating the intensity and structure of various livelihood capitals as the root causes of changes in livelihood status, while positioning the impact of national parks as the external direct causes driving transformations across these livelihood capitals.

Since livelihood, state, human capital, social capital, natural capital, physical capital, and financial capital are abstract concepts that require further refinement into measurable indicators, this study breaks down and elaborates on these concepts [67]. The system of variables for evaluating the livelihood path is shown in Tables 1 and 2.

Table 1. Variables system of the SEM for evaluating the livelihood path.

Categories	Variable	Measurement Method
National Park Impact (NPI)	Policy Impact (NPI-1)	−2 = Significant Negative Impact; −1 = Moderate Negative Impact; 0 = No Impact; 1= Moderate Positive Impact; 2 = Significant Positive Impact
	Wild Animal Impact (NPI-2)	−2 = Significant Negative Impact; −1 = Moderate Negative Impact; 0 = No Impact; 1= Moderate Positive Impact; 2 = Significant Positive Impact
Human Capital (HC)	Health Level (HC-1)	0 = Has severe illness or disability, unable to work; 1 = Has some illness or injury, can engage in limited work; 2 = No severe illness or injury, can engage in most physical labor.
	Education Level (HC-2)	0 = Illiterate; 1 = Primary school; 2 = Junior high school; 3 = High school/Vocational school; 4 = Bachelor's degree; 5 = Graduate degree or above.
	Vocational Skill level (HC-3)	0 = None; 1 = Farming, animal husbandry, special industries, new media operation, etc. (Add 1 point for each skill or experience possessed).
Social Capital (SC)	The Number of Local Relatives and Friends (SC-1)	0 = None; 1 = 1 to 5 people; 2 = 6 to 15 people; 3 = 16 to 30 people; 4 = 31 to 50 people; 5 = More than 50 people.
	The Relationship of Local Relatives and Friends (SC-2)	0 = Poor or average relationship; 1 = Good relationship; 2 = Very good relationship.
	Local Employment Opportunities (SC-3)	0 = None; 1 = Limited; 2 = Relatively Sufficient.
Natural Capital (NC)	Livestock Rearing Area (NC-1)	Calculated in "mu". *(1 mu = 666.67 m^2.)*
	Forest Land Area (NC-2)	Calculated in "mu". *(1 mu = 666.67 m^2.)*
	Cultivated Land Area (NC-3)	Calculated in "mu". *(1 mu = 666.67 m^2.)*
Physical Capital (PC)	The Quality of the House (PC-1)	1 =Earth-Timber Structure; 2 =Brick-Timber Structure; 3 =Brick-Concrete Structure; 4 = Reinforced Concrete Structure. (If multiple houses are owned, select the structurally superior one)
	Number of Production Tools (PC-2)	0 = None; 1 = 1–10 items; 2 = More than 10 items.
	Number of Durable Consumer Goods (PC-3)	0 = None; 1 = 1–5 items; 2 = More than 5 items.
Financial Capital (FC)	Borrowing Capacity (FC-1)	0 = None; 1 = Available but with a low amount; 2 = Available and with a high amount.
	Savings Balance (FC-2)	0 = None; 1 = Has savings but with a low amount; 2 = Has savings and with a high amount.
	Transfer Payment (FC-3)	0 = None; 1 = Available but with a low amount; 2 = Available with a high amount.
Livelihood State (LS)	Income (LS-1)	1 = Less than 10,000 yuan; 2 = 10,000–30,000 yuan; 3 = 30,000–50,000 yuan; 4 = 50,000–100,000 yuan; 5 = More than 100,000 yuan.
	Balance of Income and Expenditure (LS-2)	0 = Cannot cover; 1 = Can cover but with little surplus; 2 = Can cover with a substantial surplus.

Table 2. Data information of the variables system in the SEM model.

Categories	Variable	Maximum Value	Minimum Value	Average Value	Standard Deviation
NPI	NPI-1	2	−2	−0.934	0.678
	NPI-2	2	−2	−1.096	0.535
HC	HC-1	2	1	1.929	0.258
	HC-2	3	2	2.865	0.342
	HC-3	4	1	2.291	0.516
SC	SC-1	4	1	2.079	0.426
	SC-2	2	0	1.295	0.518
	SC-3	2	1	1.395	0.489
NC	NC-1	2.1	0.8	1.309	0.260
	NC-2	1.6	0.3	0.533	0.308
	NC-3	0.6	0	0.342	0.074
PC	PC-1	4	1	3.081	0.433
	PC-2	2	0	0.267	0.528
	PC-3	3	1	1.720	0.473
FC	FC-1	2	0	1.237	0.442
	FC-2	2	0	1.355	0.494
	FC-3	2	0	1.107	0.380
LS	LS-1	4	2	2.226	0.436
	LS-2	2	1	1.263	0.440

3.3. Research Methodology

This study employs structural equation modeling (SEM) as the primary quantitative analysis tool. SEM enables the exploration of mathematical relationships between latent variables that cannot be directly measured, thereby simulating correlations among multiple factors [68]. However, it is critical to note that the causal relationships between latent variables and observed variables in SEM can be either reflective or formative:

Reflective models posit latent variables as the cause of observed variables.

Formative models posit observed variables as the cause of latent variables.

Errors in model specification (e.g., misclassifying reflective/formative relationships) may lead to unreliable estimates and inaccurate descriptions of relationships between latent variables, ultimately compromising the validity of the entire study. In this study, the defined relationships between measurement variables and latent variables are shown as formative measurement models [69]. The causal relationships of latent variables under this hybrid model structure are mathematically expressed as the following:

$$\eta = \gamma_1 X_1 + \gamma_2 X_2 + \cdots + \gamma_n X_n + \delta$$

η represents the latent variable;

X_n denotes the n-th observed variable;

γ_n is the path coefficient of the n-th variable;

δ is the residual error term.

This equation indicates that a latent variable η is formed by the aggregation of n observed variables ($X_1, X_2, \ldots X_n$). The explanatory power of each observed variable X_n on the latent variable η is quantified by γ_n, while δ accounts for unavoidable measurement error. Although the error term δ is typically not displayed, given that structural equation modeling (SEM) relies on subjective perceptions as research data, the inclusion of error terms can ensure the reliability of the study.

Integrating the Sustainable Livelihoods Framework (SLF) defined earlier, the research model can be formally operationalized within the SEM framework. Specifically, the hypothesized relationships among livelihood capitals (e.g., social, natural) and livelihood outcomes are mapped onto the structural and measurement components of the SEM equations, enabling a systematic test of the theoretical propositions.

Given that the measurement relationships between latent variables and observed variables in this study's design are shown as formative indicators, the use of SmartPLS 3.0 software (a software developed by SmartPLS GmbH, working as a data analysis tool that uses Partial Least Squares (PLS) path modeling to estimate structural equation models) demonstrates high methodological suitability for data analysis in this study. SmartPLS 3.0 specializes in partial least squares structural equation modeling (PLS-SEM), a methodology particularly suitable for small sample sizes, non-normally distributed data, and complex models (e.g., formative measurement models) [70]. The key advantages of its algorithm include a high tolerance for multicollinearity and a focus on predictive power over model fit indices [71]. By enabling the deconstruction of structural equations built upon formative indicator systems, SmartPLS 3.0 is adept for exploratory research in social sciences [72]. After combining with the variables system, the SEM model is illustrated in Figure 3.

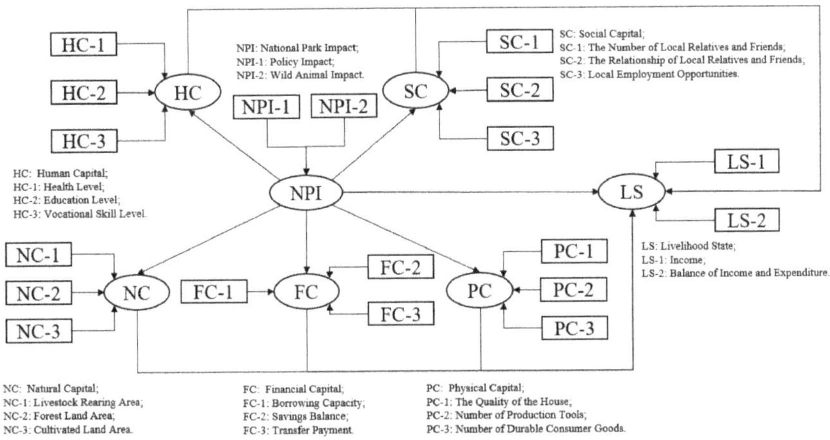

Figure 3. The structure of the SEM model for evaluating the livelihood path.

This study has designed the relationships among national park impacts, livelihood capitals, and livelihood state, but in the specific structural equation, the three hypotheses (H1, H2, H3) need to be refined. When using structural equation modeling (SEM) to investigate relationships between latent variables, it is necessary to formulate hypotheses regarding the relationships among these latent variables and evaluate whether these hypotheses are empirically supported through data analysis [73]. This process allows for a discussion of the model's explanatory power. As this study fundamentally follows an exploratory research logic grounded in field survey data, it cannot pre-estimate the statistical significance of effects (i.e., the magnitude of path coefficients between variables). However, reasonable hypotheses about the directionality (positive/negative nature) of these relationships can be proposed, as follows:

H1-1. *Human capital exerts a positive influence on livelihood states;*

H1-2. *Social capital exerts a positive influence on livelihood states;*

H1-3. *Natural capital exerts a positive influence on livelihood states;*

H1-4. *Physical capital exerts a positive influence on livelihood states;*

H1-5. *Financial capital exerts a positive influence on livelihood states;*

H2-1. *National park impact exerts a negative influence on human capital;*

H2-2. *National park impact exerts a negative influence on social capital;*

H2-3. *National park impact exerts a negative influence on natural capital;*

H2-4. *National park impact exerts a negative influence on physical capital;*

H2-5. *National park impact exerts a negative influence on financial capital;*

H3. *National park impact exerts a negative influence on livelihood state.*

4. Results

4.1. Model Validation

In the process of structural equation modeling (SEM) analysis, it is essential to validate the reliability and validity of field survey data. Given that all constructs in this study's model were mostly designed as formative indicators, SmartPLS 3.0 could not perform conventional reliability and validity tests (e.g., Cronbach's alpha, KMO measure) [74]. Therefore, the analysis shifted to SPSS 19.0 statistical software (a professional software developed by International Business Machines Corporation, which can be used to analyze the correlations between data.) to calculate Cronbach's alpha coefficients and the Kaiser–Meyer–Olkin (KMO) measure, thereby assessing whether the data met required reliability and validity standards. It was also necessary to address potential multicollinearity issues; SmartPLS 3.0 was utilized to conduct collinearity diagnostics by computing variance inflation factors (VIFs).

This study assessed the reliability of the questionnaire data using Cronbach's alpha coefficient. Generally, a Cronbach's alpha coefficient above 0.7 is considered good, between 0.4 and 0.7 is moderate, and below 0.4 is poor [75]. The Cronbach's alpha coefficients of the questionnaire items in this study met the basic threshold for exploratory research, indicating that the data exhibited adequate reliability (Table 3).

Table 3. Results of questionnaire reliability measurement in the SEM structure.

Latent Variables	Cronbach's Alpha	Items
NPI	0.734	2
HC	0.705	3
SC	0.688	3
NC	0.739	3
PC	0.628	3
FC	0.501	3
LS	0.718	2

To validate the construct validity of the questionnaire data in this study, a factor analysis was conducted. The Kaiser–Meyer–Olkin (KMO) measure and Bartlett's test of sphericity were employed to assess the structural validity. Generally, the KMO value should exceed 0.5 [76]; in this study, the KMO coefficient was 0.636, meeting the threshold. Bartlett's test requires the data distribution to exhibit sphericity with a significance probability (p-value) below 0.05 [77]. The test results in this study also satisfied these criteria

($p < 0.05$). Therefore, the questionnaire data demonstrated adequate construct validity and was suitable for further in-depth analysis.

In SmartPLS analyses, formative indicators are not suitable for calculating composite reliability (CR) or average variance extracted (AVE). Instead, collinearity diagnostics must be conducted, typically using the variance inflation factor (VIF) as the criterion, with VIF values required to be below 3.3 [78]. Both the outer model (latent variable formative measurement model) and the inner model (latent variable structural relationship model) in this study satisfied this requirement (Tables 4 and 5).

Table 4. Multicollinearity test results (VIF values) for inner model in the SEM structure.

Latent Variables	Observed Variables	VIF
NPI	NPI-1	1.148
	NPI-2	1.178
HC	HC-1	1.093
	HC-2	1.122
	HC-3	1.063
SC	SC-1	1.013
	SC-2	1.002
	SC-3	1.015
NC	NC-1	1.112
	NC-2	2.070
	NC-3	2.229
PC	PC-1	1.170
	PC-2	1.206
	PC-3	1.071
FC	FC-1	1.388
	FC-2	1.347
	FC-3	1.046
LS	LS-1	1.215
	LS-2	1.233

Table 5. Multicollinearity test results (VIF values) for outer model in the SEM structure.

Dependent Variables	Independent Variables	VIF
HC	NPI	1.203
SC		1.158
NC		1.632
PC		1.337
FC		1.272
LS	HC	1.094
	SC	1.115
	NC	1.621
	PC	1.268
	FC	1.350

4.2. Hypothesis Testing

The significance of path relationships between variables was tested using the bootstrap method in SmartPLS3.0, with 5000 bootstrap resamples generated from the original sample size of 567. All path coefficients (Tables 6 and 7) in both the structural model (inner model) and measurement model (outer model) were statistically significant ($p < 0.05$).

Table 6. Hypothesis testing of the structural model in the SEM structure.

Hypothesized Path	*t*-Statistics	*p*-Value	Decision
NPI > HC	3.132	0.025	Supported
NPI > SC	3.268	0.024	Supported
NPI > NC	6.822	0.011	Supported
NPI > PC	4.384	0.013	Supported
NPI > FC	4.128	0.007	Supported
HC > LS	2.363	0.018	Supported
SC > LS	3.165	0.013	Supported
NC > LS	6.105	0.012	Supported
PC > LS	2.457	0.014	Supported
FC > LS	3.503	0.021	Supported

Table 7. Hypothesis testing of the measurement model in the SEM structure.

Hypothesized Path	*t*-Statistics	*p*-Value	Decision
NPI-1 > NPI	2.532	0.011	Supported
NPI-2 > NPI	5.762	0.018	Supported
HC-1 > HC	2.895	0.013	Supported
HC-2 > HC	2.144	0.025	Supported
HC-3 > HC	9.747	0.012	Supported
SC-1 > SC	3.238	0.017	Supported
SC-2 > SC	2.069	0.039	Supported
SC-3 > SC	5.997	0.022	Supported
NC-1 > NC	6.083	0.017	Supported
NC-2 > NC	7.620	0.028	Supported
NC-3 > NC	22.701	0.033	Supported
PC-1 > PC	3.478	0.041	Supported
PC-2 > PC	25.083	0.012	Supported
PC-3 > PC	2.673	0.018	Supported
FC-1 > FC	3.272	0.031	Supported
FC-2 > FC	18.321	0.022	Supported
FC-3 > FC	3.462	0.029	Supported
LS > LS-1	14.508	0.013	Supported
LS > LS-2	11.071	0.021	Supported

The findings demonstrate that the hypothesized pathway relationships proposed in this study exhibited exceptionally high statistical significance. The direction (positive/negative) and magnitude of the path coefficients accurately reflected the relationships between latent variables, lending themselves to robust further investigation and scholarly discourse.

4.3. Path Coefficient of the Livelihood

The path coefficients (Table 8) derived from the SmartPLS 3.0 analysis confirmed the hypothesized relationships among variables in the structural equation model, thereby validating the theoretical assumptions of this study.

Table 8. Path coefficients of the SEM model for evaluating the livelihood path.

Effect	Path	Value
Direct Effect	NPI > HC	−0.219
	NPI > SC	−0.217
	NPI > NC	−0.383
	NPI > PC	−0.298
	NPI > FC	−0.258
	HC > LS	0.120
	SC > LS	0.225
	NC > LS	0.351
	PC > LS	0.164
	FC > LS	0.179
Indirect Effect	NPI > HC > LS	−0.026
	NPI > SC > LS	−0.049
	NPI > NC > LS	−0.134
	NPI > PC > LS	−0.049
	NPI> FC > LS	−0.046
Total Effect	NPI > LS	−0.256

In the structural model of the livelihood system, the strongest positive relationship was observed in the "NC → LS" path (natural capital → livelihood status), with a path coefficient of 0.351. Other livelihood capitals (human, social, physical, financial) also exhibited positive effects on livelihood status. These results imply that for the study's target population, national park environmental resettlements, natural capital serves as the most critical core livelihood capital and the primary driver of household income within their livelihood system. The impact of different livelihood capitals on livelihood state varied significantly. Specifically, a one standard-deviation (SD) increase in human capital, social capital, natural capital, physical capital, and financial capital corresponded to increases of 0.120 SD, 0.225 SD, 0.351 SD, 0.164 SD, and 0.179 SD in livelihood status, respectively. The findings above validate Hypothesis H1, demonstrating that enhancements of livelihood capital in any form contribute to improved livelihood state.

In the structural model of national park impact, the path with the strongest negative impact relationship was "NPI → NC" (national park impact on natural capital), demonstrating a coefficient of −0.383. Furthermore, the national park's effects on other livelihood capitals also exhibited negative tendencies, validating Hypothesis H2. These results indicate that the establishment of national parks, as an exogenous social event, not only diminishes residents' livelihood capital through exclusionary policies but also generates persistent adverse impacts on human communities' production and daily life by continuously expanding wildlife populations.

Within the formation structures of social capital, natural capital, physical capital, and financial capital, each contained a single dominant core indicator. Specifically, HC-3 demonstrated the strongest influence ($\beta = 0.713$) in determining human capital levels; SC-3 ($\beta = 0.502$) served as the primary determinant of social capital; NC-1 ($\beta = 0.685$) exerted the strongest effect on natural capital; PC-2 ($\beta = 0.802$) dominated physical capital formation; and FC-2 ($\beta = 0.784$) emerged as the key driver of financial capital. Notably, human capital exhibited the most balanced internal formation pattern. Overall, all measurement indicators in the five livelihood capital models demonstrated positive contributions, albeit with varying magnitudes of influence.

The effects of LS-1 and LS-2 on LS were 0.856 and 0.827, respectively, indicating strong positive influences. This finding suggests that higher levels of livelihood status intensity correlate with increased income levels and an enhanced ability to maintain balanced budgets among national park residents.

Based on the results of the structural model and measurement model, NC-3 (Cultivated Land Area) was the core indicator that determined livelihood status. Nature capital, which was mainly contributed to by NC-3, plays a central driving role in the development of the livelihood capacity of residents and is the main source that ultimately affects the income level and financial balance of these residents. Empirical analysis revealed that the pathway 'NPI > NC > LS' exhibited the strongest indirect effect ($\beta = -0.134$), indicating its dominant mediating role in the observed livelihood dynamics. Similarly, due to the most severe impact on natural capital resulting from the construction of national parks, the overall livelihood capacity of local residents will also be reduced.

Finally, as a comparative illustration, the total effect of national parks on residents' livelihood state was quantified as $\beta = -0.256$. This indicates that for every 1-standard-deviation (SD) increase in policy exclusion and wildlife-induced disturbances originating from national parks, residents' livelihood status decreased by 0.256 SD. These results validate Hypothesis H3, demonstrating that national parks—as external entities with dual natural and social attributes—exert a significant adverse impact on the maintenance and improvement of local livelihoods.

In this study, Figure 4 is specifically developed to provide a visual representation of the complex relationships among national park impacts, residents' livelihood capital, and livelihood state.

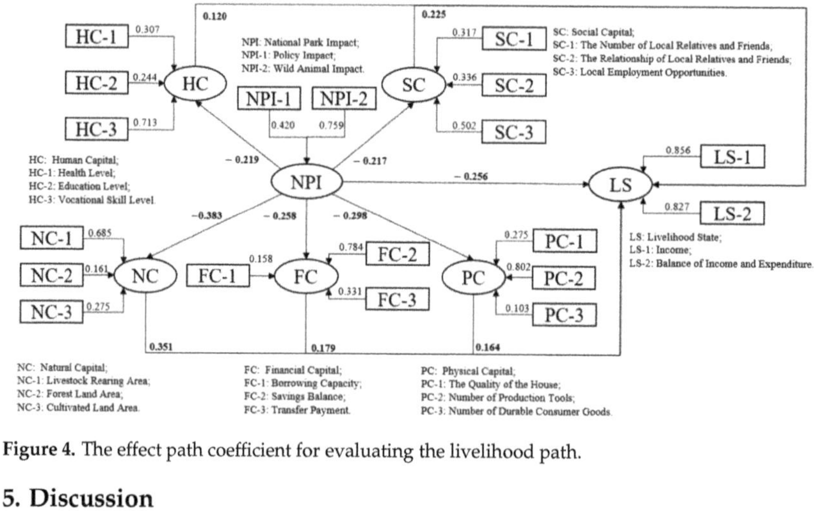

Figure 4. The effect path coefficient for evaluating the livelihood path.

5. Discussion

The research findings clearly demonstrate that the pathways through which national park establishment impacts local residents' livelihoods manifest in two primary dimensions: restrictive development policies and expanding wildlife populations, which jointly erode the residents' livelihood capital system. This cumulative effect ultimately diminishes overall livelihood capacity, shown as income reduction and fiscal imbalance. Quantitative analyses substantiate the hypotheses proposed in this exploratory study, revealing the precise mechanisms by which external social events disrupt livelihood systems. Consequently, it becomes imperative to systematically examine the impact mechanisms behind livelihood deterioration, assess emerging livelihood risks, and formulate viable improvement strategies for affected communities.

5.1. Impacts of National Parks on Residents' Livelihood

The impacts of national park establishment on local residents' production and living patterns can be systematically categorized into two dimensions: (1) land-use restrictions inherent to park development [31] and (2) the increased frequency of human–wildlife conflicts following wildlife population growth [79]. These dual pressures usually degrade various livelihood capitals, ultimately elevating systemic livelihood vulnerability [80]. The specific manifestations include the following:

The exclusionary nature of national park governance triggers the withdrawal of infrastructure and social services, resulting in constrained access to educational and medical resources [81]. This institutional marginalization leads to irreversible human capital depletion through deteriorating health conditions and educational attainment [82]. Concurrently, escalating wildlife encroachment poses direct threats to personal safety, reflecting an inherent ecological–social paradox between conservation objectives and human settlement needs.

Beyond individual hardships, national park implementation generates community disintegration risks [83]. As human communities fundamentally rely on shared production–living systems, the spatial overlap between protected areas and human communities creates inevitable tension. The dual pressures of regulatory constraints and ecological displacement progressively erode social capital: the vital bonding networks that sustain collective livelihood strategies.

The most acute manifestation of national park-induced exclusion lies in human–wildlife competition for land resources, rooted in the dual nature of land tenure [84]. Geospatial overlap renders land as both the foundational requirement for park establishment and the core productive asset for community livelihoods, necessitating stringent regulatory constraints on land use. Simultaneously, wildlife encroachment extends beyond physical threats to also systematically undermine agricultural productivity through crop destruction and asset damage. Thus, national parks operationalize natural capital depletion through twin mechanisms: policy-imposed land restrictions and ecological restoration-driven wildlife population growth [85].

National park establishment fundamentally disrupts the material basis of sustainable living [86]. Housing quality, productive equipment stocks, and durable household assets—key components of material capital—lose functional relevance under constrained livelihood systems. This systemic depreciation directly diminishes comprehensive livelihood capacity.

The exclusionary imperative of national park governance forces residents to deplete savings buffers to compensate for other capital losses. Concurrently, diminished livelihood capacity impairs creditworthiness for formal financing [87]. Coupled with inadequate fiscal transfer mechanisms in underdeveloped regions, these dynamics inevitably degrade financial capital robustness within livelihood portfolios.

Compared to the inherent exclusion of human activities and communities in national park establishment itself, the increased frequency and intensity of wildlife incidents resulting from park development have exerted greater and more persistent negative impacts on local residents' livelihoods. Moreover, such impacts cause more extensive damage to various livelihood capitals. Previous studies predominantly focused on singular aspects of this issue and overlooked the dual social and natural attributes intrinsic to national parks. In reality, resolving conflicts between national park management and residents' livelihood security necessitates heightened attention to wildlife-induced disturbances [79,82].

5.2. Risk Management

Based on the preceding analysis, national parks have jointly impaired residents' livelihood capacity through two dimensions: policy restrictions and wildlife impacts. This indicates that risk management should be addressed from both the national park and livelihood capital perspectives.

Regarding national park construction, it is essential to reasonably control the exclusion of human activities within park boundaries. For residents who cannot relocate promptly, basic livelihood safeguards must be implemented, particularly concerning land as a critical means of production [88]. A portion of essential land should be retained for residents to sustain their fundamental productive and daily needs. Simultaneously, given the rising wildlife populations, it is imperative to refine compensation systems for wildlife-related incidents, appropriately increase compensation amounts, streamline administrative procedures, and integrate external commercial insurance to strengthen protections for residents' personal and property safety [4,89,90].

Regarding risks to livelihood capital, the ecological compensation system must be enhanced. Comprehensive investigation and registration mechanisms should be established to accurately assess residents' losses, with dedicated funding accounts created for financial redress. Additionally, livelihood recovery for residents should be prioritized and systematically integrated into the planning and implementation of national park development. This approach aims to achieve a dual sustainable development, balancing ecological conservation with the socioeconomic well-being of local communities [91–93].

Concurrently, it is noteworthy that the Northeast Tiger and Leopard National Park is situated within a transboundary region encompassing China, Russia, and North Korea. While China and Russia have established preliminary cooperation concerning the joint conservation of wildlife within their contiguous border areas, such collaboration remains predominantly research-oriented at present [1]. Notably, a unified fund for the compensation of affected residents has yet to be established and managed bilaterally; instead, respective national authorities provide compensatory relief independently to their impacted residents [11]. Effective communication channels with North Korea regarding collaborative biodiversity conservation and economic coordination remain undeveloped [19]. Domestically, within China, the current level of human–wildlife conflict and associated economic compensation remains manageable, largely attributable to the low population density in the park's vicinity. However, anticipating future growth trends in wildlife populations, the establishment of a more efficient and responsive compensation system for wildlife-inflicted damage proves imperative.

5.3. Livelihood Improvement

The majority of residents within national park construction zones predominantly rely on land-based economic production as their primary source of household income, occasionally supplementing through migratory labor opportunities [35,94]. Given that land, as the most critical productive asset, is now subject to development restrictions, this livelihood model has demonstrably lost its sustainability [95]. Consequently, implementing targeted interventions to facilitate the transition of livelihood sustenance mechanisms has become imperative. Based on these situations, livelihood transition strategies should be implemented with two strategies:

Maintain existing livelihood frameworks while enhancing production technologies and augmenting reproduction funds to circumvent land scarcity constraints, thereby boosting agricultural economic returns. This strategy could be marked as an "Intensification Approach".

Transform traditional livelihood systems by redirecting household labor and time allocations toward non-primary industrial activities, achieving capacity enhancement through structural transition. This strategy could be marked as a "Diversification Approach".

Whether these livelihood enhancement strategies can be effectively implemented is contingent on both generating sufficient agency at the individual level of residents and the provision of essential support from the government, businesses, and other social organizations.

It should also be recognized that national park construction recruits a significant number of local residents as forest rangers. This provision of jobs can somewhat compensate for local residents' economic losses and promote changes and improvements in their livelihood patterns. The government and scholars should pay attention to such positive opportunities.

6. Conclusions

This study constructed a livelihood capital composition system for residents in national parks and employed structural equation modeling as the primary analytical tool to examine the impact of various livelihood capitals on the ultimate livelihood levels of residents. Key findings revealed that natural capital exerted the most significant influence on the livelihood capital system of residents, followed by social capital, financial capital, physical capital, and human capital. Among these, cultivated land area emerged as the most impactful factor within the livelihood capital framework.

Aligning with these findings and the master plan of the Northeast Tiger and Leopard National Park, three critical insights emerge. First, maintaining essential land resources proves crucial, as agricultural production remains the primary livelihood source for residents. Land resources serve as irreplaceable production factors; the substantial loss of land would render additional labor and capital investments meaningless. Second, industrial transformation requires phased implementation. Gradual restrictions on land development could provide temporal flexibility for stable livelihood transition. Third, the master plan essentially seeks equilibrium between national park construction and local residents' livelihoods while pursuing sustainable development. Although inherent contradictions exist (e.g., land use conflicts), targeted policies could enhance the feasibility of sustainable development.

It should be noted that national park construction can enhance ecological benefits and effectively protect cultural assets. In addition to addressing the impacts on local livelihoods, attention should also be paid to the development opportunities which could be achieved by local residents. A more rigorous method is also needed to quantify the gap between residents' actual behavior and their stated expectations. However, due to the limitations of the length and subject of this study, this issue will be further discussed in the future.

Based on these conclusions, this study proposes the following policy recommendations to maintain basic livelihood security for residents: (1) Permit residents to retain a certain amount of arable land and appropriately extend policies aimed at safeguarding their livelihoods; (2) Provide residents with acceptable vocational skills training and promote the adoption of scientific cultivation techniques to enhance per-unit-area yield; (3) Offer residents access to necessary credit support, thereby ensuring financial conditions conducive to the stable advancement of agricultural reproduction; (4) Actively expand alternative employment opportunities, aiming to transition as many farmers as possible into wage laborers within a five-year period, so as to mitigate the constraints on agricultural production imposed by limited land availability through employment absorption in the secondary and tertiary sectors. (5) Improve the wildlife-caused damage compensation mechanism by moderately increasing compensation amounts and simplifying administrative procedures, thereby providing timely and adequate social compensation for potential losses of life or property suffered by residents.

Author Contributions: Conceptualization, L.G. and G.S.; methodology, L.G.; software, Y.Z.; validation, L.G., H.L. and X.Y.; formal analysis, L.G.; investigation, L.G., Y.Z., H.L. and X.Y.; resources, G.S.; data curation, G.S.; writing—original draft preparation, L.G.; writing—review and editing, L.G. and G.S.; visualization, L.G.; supervision, G.S.; project administration, G.S.; funding acquisition, G.S. All authors have read and agreed to the published version of the manuscript.

Funding: This research was funded by the China Ministry of Education Research Program (Fund 22YJCZH036).

Data Availability Statement: The original contributions presented in the study are included in the article; further inquiries can be directed to the corresponding author.

Acknowledgments: During the investigation and writing of this study, assistance was received from the forestry departments of Jilin and Heilongjiang Provinces in China.

Conflicts of Interest: The authors declare no conflicts of interest.

References

1. Zhao, W. Beginning: China's national park system. *Natl. Sci. Rev.* **2022**, *9*, nwac150. [CrossRef] [PubMed]
2. Tian, M.; Fang, S. A Review of National Park Selection Criteria System in China: A Case Study of 9 National Park Experimental Areas. *World For. Res.* **2017**, *30*, 62–68.
3. Belay, S.; Amsalu, A.; Abebe, E. Awash National Park, Ethiopia: Use policy, ethnic conflict and sustainable resources conservation in the context of decentralization. *Afr. J. Ecol.* **2013**, *51*, 122–129. [CrossRef]
4. Emphandhu, D.Z. *Land Use Conflict and Public Participation as a Conflict Management Tool in Park Management Planning Process: A Case Study of Doi Inthanon National Park in Thailand*; University of Washington: Seattle, WA, USA, 1992.
5. Marquardt, M.A. Settlement and Resettlement—Experience from Uganda National-Parks and Reserves. In *The Involuntary Resettlement in Africa: Selected Papers from A Conference on Environment and Settlement Issues in Africa*; World Bank Publications: Washington, DC, USA, 1994; pp. 147–159.
6. Schmidt-Soltau, K. Conservation-related resettlement in Central Africa: Environmental and social risks. *Dev. Change* **2003**, *34*, 525–551. [CrossRef]
7. Cernea, M.M.; Schmidt-Soltau, K. Corrigendum to "Poverty Risks and National Parks: Policy Issues in Conservation and Resettlement" [World Development 34 (2006) 1808–1830]. *World Dev.* **2008**, *36*, 2966. [CrossRef]
8. Neelakantan, A.Z. Moving People for Tigers: Resettlement, Food Security and Landscape-Level Conservation in Central India. Ph.D. Thesis, Columbia University, New York, NY, USA, 2019.
9. Andrade, G.S.M.; Rhodes, J.R. Protected Areas and Local Communities: An Inevitable Partnership toward Successful Conservation Strategies? *Ecol. Soc.* **2012**, *17*, 14. [CrossRef]
10. He, Q.; Chen, J.; Li, H. Experience of Canadian Grasslands National Park in Construction and Management and Its Enlightenment. *World For. Res.* **2023**, *36*, 105–110.
11. Wang, J.-H.Z. National parks in China: Parks for people or for the nation? *Land Use Policy* **2019**, *81*, 825–833. [CrossRef]
12. Zhou, W.; Chen, K. Community Co-Management Model in New Zealand National Parks: Experience and Enlightenment. *World For. Res.* **2024**, *37*, 105–110.
13. Chen, L.; Wu, W.; Wang, G. Evolution and Enlightenment of the Interaction Between Canadian National Parks and Indigenous People. *World For. Res.* **2021**, *34*, 98–103.
14. Sneed, P.G.Z. National Parklands in Northern Homelands: A Comparison of Co-Management of National Parks with Native People in Alaska (United States of America) and the Yukon (Canada). Ph.D. Thesis, University of Hawai'I, Honolulu, HI, USA, 1998.
15. He, S.; Su, Y.; Cheng, H. Coordinating community resource use and conservation: An institutional diagnostic practice in the Wuyishan National Park. *J. Environ. Manag.* **2022**, *317*, 115508. [CrossRef] [PubMed]
16. He, S.; Wei, Y. Measuring the conservation attitudes of local farmers towards conservation easements in the Qianjiangyuan National Park. *Glob. Ecol. Conserv.* **2022**, *36*, e02123. [CrossRef]
17. Dewi, S.; Belcher, B.; Puntodewo, A. Village economic opportunity, forest dependence, and rural livelihoods in East Kalimantan, Indonesia. *World Dev.* **2005**, *33*, 1419–1434. [CrossRef]

18. Xu, J.; Huang, J.; Li, X.; Ning, Y. Review and Discussion on China's Strategies of National Park System Development. *World For. Res.* **2017**, *30*, 58–62.
19. Zhang, G.; Qu, Z. A Review of Domestic Research and Practices on National Parks in China. *World For. Res.* **2019**, *32*, 57–61.
20. Lu, Z. National parks in China: A challenge or an opportunity? *Biodivers. Sci.* **2014**, *22*, 421–422. [CrossRef]
21. Shi, J.; Huang, Y. Research Progress on National Park Management. *World For. Res.* **2019**, *32*, 40–44.
22. Duan, T.; Li, N.; Huang, Z.; Li, Y.; Mu, Y.; Xiao, W. Progress of China's national park construction and development prospects. *Acta Ecol. Sin.* **2024**, *44*, 4964–4972.
23. Yang, S.; Xu, H.; Zhao, Z.; Wang, F.; Sun, Q.; Tang, J.; He, J.; Zhong, L. Greenhouse gas emissions in China's national Parks: A 2020 National-Scale analysis and Implications for management. *Ecol. Indic.* **2025**, *170*, 113089. [CrossRef]
24. Zhang, Y.; Vanclay, F.; Hanna, P. How communities and social impacts are considered in policies for protected areas in China. *Land Use Policy* **2025**, *148*, 107404. [CrossRef]
25. Wang, X.; Lopez-Carr, D.; Zhou, L. Combined Effects of Forest Conservation and Population Resettlement on the Ecological Restoration of Qilian Mountain National Park. *Land* **2024**, *13*, 1983. [CrossRef]
26. Wang, Q.; Wu, Y.; Ren, M.; Li, B.L. Suggestions on optimizing the planning of national parks and national nature reserves in China. *Sci. Technol. Rev.* **2023**, *41*, 14–20.
27. Zhao, X.; Su, H. The research framework and key issues of sustainable livelihoods in the national park. *J. Nat. Resour.* **2023**, *38*, 2217–2236. [CrossRef]
28. Wang, Y.; Hu, Y.; Liu, J. Overview of China's national park system reform process. *Chin. J. Popul. Resour. Environ.* **2024**, *22*, 481–492. [CrossRef]
29. Kang, X.; Du, M.; Zhao, L.; Liu, Q.; Liao, Z.; Su, H.; Xiang, T.; Gou, C.; Liu, N. Integrity-centered framework for determining protected areas boundary: An application in the China's national park. *Ecol. Inform.* **2024**, *84*, 102885. [CrossRef]
30. Ma, W.; Hou, S.; Su, W.; Mao, T.; Wang, X.; Liang, T. Estimation of carbon stock and economic value of Sanjiangyuan National Park, China. *Ecol. Indic.* **2024**, *169*, 112856. [CrossRef]
31. Xu, M.; Ma, Z.; Gao, J.; Li, M. The impact of farmers' participation in national park construction on household income in China: Evidence from the Qinling Area of the Giant Panda National Park. *J. Environ. Manag.* **2025**, *386*, 125743. [CrossRef] [PubMed]
32. Yu, H.; Hu, X. Potential construction area identification of the transboundary national park bridging ecology, society and economics: A case study of Mount Everest region. *J. Environ. Manag.* **2025**, *381*, 125190. [CrossRef] [PubMed]
33. Zhao, X.; Chen, Q.; Chen, T. Human—Animal conflicts and the cognition of herdsmen in the Three—River—Source National Park. *J. Arid Land Resour. Environ.* **2022**, *36*, 39–46.
34. Cheng, Q.; Cheng, X.; Ma, K.; Zhao, X.; Qu, J. Offering the win-win solutions between ecological conservation and livelihood development: National parks in Qinghai, China. *Geogr. Sustain.* **2020**, *1*, 251–255. [CrossRef]
35. Yu, P.; Zhang, J.; Wang, Y.; Wang, C.; Zhang, H. Can tourism development enhance livelihood capitals of rural households? Evidence from Huangshan National Park adjacent communities, China. *Sci. Total Environ.* **2020**, *748*, 141099. [CrossRef] [PubMed]
36. Amery, J.; Philpott, A. Undernutrition Under Attention: The Changing Approach of the UK Department for International Development to Nutrition. *IDS Bull. Inst. Dev. Stud.* **2009**, *40*, 125–129. [CrossRef]
37. Treffgarne, C.B.W. Joined-up government? Insights from education during DFID's first decade. *Int. J. Educ. Dev.* **2019**, *68*, 45–55. [CrossRef]
38. Guo, A.; Wei, Y.; Zhong, F.; Wang, P. How do climate change perception and value cognition affect farmers' sustainable livelihood capacity? An analysis based on an improved DFID sustainable livelihood framework. *Sustain. Prod. Consum.* **2022**, *33*, 636–650. [CrossRef]
39. Delina, L.L.; Gaviola, J.H.; Cagoco-Guiam, R. Climate and security risks and their implications for sustainable livelihoods: The case of Maguindanao in conflict-ridden Philippine Bangsamoro. *World Dev. Perspect.* **2024**, *36*, 100642. [CrossRef]
40. Sultana, M.; Shetu, J.F.; Sarker, F.; Ahammad, S.; Sujan, M.H.K. How do improved lentil varieties affect farmers' livelihood status in central Bangladesh? *Heliyon* **2025**, *11*, e41189. [CrossRef] [PubMed]
41. Lun, Y.; Na, Q.; Ning-tao, T. The livelihood-resource nexus in UNESCO biosphere reserves: A case study from Chebaling Biosphere Reserve in China. *J. Nat. Conserv.* **2025**, *84*, 126802. [CrossRef]
42. Han, W.; Zhang, Z.; Zhang, X. How does household farmland rental behavior affect gender differences in labor division and livelihood strategy? Insights from the household production theory. *Land Use Policy* **2024**, *147*, 107362. [CrossRef]
43. Onyeyirichi, O.A.; Deepika, M.G. Rural multidimensional poverty and livelihood mix: A micro level study in Bihar, India. *Heliyon* **2025**, *11*, e42772. [CrossRef] [PubMed]
44. Hussain, A.; Mandić, A.; Fusté-Forné, F. Transforming communities: Analyzing the effects of infrastructure and tourism development on social capital, livelihoods, and resilience in Gilgit-Baltistan, Pakistan. *J. Hosp. Tour. Manag.* **2024**, *59*, 276–295. [CrossRef]
45. He, X.; Zhong, R.; Zhao, W.; Sun, L.; Sun, R. Livelihood changes of migrants due to hydropower construction: A case study from Southwestern China. *Appl. Geogr.* **2025**, *176*, 103511. [CrossRef]

46. Rahman, M.S.; Majumder, M.K.; Sujan, M.H.K.; Manjira, S. Livelihood status of coastal shrimp farmers in Bangladesh: Comparison before and during COVID-19. *Aquac. Rep.* **2021**, *21*, 100895. [CrossRef]
47. Shanazi, K.; Aazami, M. Sustainability and diversity of Wetland livelihoods: The case of livelihoods around the Zarivar Wetland, Iran. *Environ. Dev.* **2025**, *55*, 101212. [CrossRef]
48. Toku, A.; Amoah, S.T.; N-yanbini, N.N.; Sarfo, A.K.; Tornyeviadzi, P. Analysis of land-use and livelihood transformations in peri-urban areas: Insights from North-Western Ghana. *Heliyon* **2024**, *10*, e38467. [CrossRef] [PubMed]
49. Liu, W.; Qi, P.; Xu, J.; Feldman, M.; Xu, D. Does livelihood risk matter in disaster preparedness? Insights from flood risk areas of rural China. *Clim. Risk Manag.* **2025**, *48*, 100705. [CrossRef]
50. Pilone, E.; Demichela, M. A semi-quantitative methodology to evaluate the main local territorial risks and their interactions. *Land Use Policy* **2018**, *77*, 143–154. [CrossRef]
51. Zhang, R.; Zhou, S.; Chen, R. Moderating effects of grassland ecological compensation policy in linking climatic risk and farmers' livelihood resilience in China. *Clim. Smart Agric.* **2025**, *2*, 100040. [CrossRef]
52. Yan, S.; Deng, X.; Li, Z. The impact of national park construction on the livelihood capital of original residents: A case study of the Qilian Mountain National Park system pilot. *Sustain. Futures* **2024**, *8*, 100380. [CrossRef]
53. Thanvisitthpon, N.; Shrestha, S.; Pal, I.; Ninsawat, S.; Chaowiwat, W. Assessment of flood adaptive capacity of urban areas in Thailand. *Environ. Impact Assess. Rev.* **2020**, *81*, 106363. [CrossRef]
54. Zhuang, L.; He, J.; Yong, Z.; Deng, X.; Xu, D. Disaster information acquisition by residents of China's earthquake-stricken areas. *Int. J. Disaster Risk Reduct.* **2020**, *51*, 101908. [CrossRef]
55. Zeng, X.; Guo, S.; Deng, X.; Zhou, W.; Xu, D. Livelihood risk and adaptation strategies of farmers in earthquake hazard threatened areas: Evidence from sichuan province, China. *Int. J. Disaster Risk Reduct.* **2021**, *53*, 101971. [CrossRef]
56. Dovie, D.B.K.; Pabi, O. Partial climatic risk screening, adaptation and livelihoods in a coastal urban area in Ghana. *Habitat Int.* **2023**, *138*, 102868. [CrossRef]
57. Galarza-Villamar, J.A.; Leeuwis, C.; Cecchi, F. Rice farmers and floods in Ecuador: The strategic role of social capital in disaster risk reduction and livelihood resilience. *Int. J. Disaster Risk Reduct.* **2024**, *104*, 104332. [CrossRef]
58. Krauss, J.E.; Castro, E.; Kingman, A.; Nuvunga, M.; Ryan, C. Understanding livelihood changes in the charcoal and baobab value chains during COVID-19 in rural Mozambique: The role of power, risk and civic-based stakeholder conventions. *Geoforum* **2023**, *140*, 103706. [CrossRef] [PubMed]
59. Zeleke, G.; Teshome, M.; Ayele, L. Farmers' livelihood vulnerability to climate-related risks in the North Wello Zone, northern Ethiopia. *Environ. Sustain. Indic.* **2023**, *17*, 100220. [CrossRef]
60. Jing, Z.; Yu, Y.; Wang, Y.; Su, X.; Qiu, X.; Yang, X.; Xu, Y. Study on the mechanism of livelihood behavior decision of rural residents in ethnic tourism villages in Western Sichuan. *Ecol. Indic.* **2024**, *166*, 112250. [CrossRef]
61. Gao, Y.; Bi, Y.; Yu, L. Understanding the impacts of ecological compensation policy on rural livelihoods: Insights from forest communities of China. *J. Environ. Manag.* **2025**, *374*, 123921. [CrossRef] [PubMed]
62. Xu, W.; Zang, Z.; Du, A.; Ouyang, Z. The experiences of Northeast China Tiger and Leopard National Park pilot. *Biodivers. Sci.* **2021**, *29*, 295–297. [CrossRef]
63. Jiang, Y.; Tian, J.; Zhao, J.; Tang, X. The connotation and assessment framework of national park ecosystem integrity:A case study of the Amur Tiger and Leopard National Park. *Biodivers. Sci.* **2021**, *29*, 1279–1287. [CrossRef]
64. Zhu, H.; Zhao, M.; Zhu, Z. Practice and Enlightenment of National Park Concession at Home and Abroad:A Case Study of Northeast China Tiger and Leopard National Park. *World For. Res.* **2022**, *35*, 50–55.
65. Gu, Y.; Zhang, F.; Liang, X.; Liu, C.; Xing, S.; Wang, Q. Integration of natural reserves based on potential habitat protection of the Amur tiger. *Chin. J. Ecol.* **2020**, *39*, 1590–1599.
66. Yang, Y.; Li, X.; Liu, Q. Coupling and Cooperative Evaluation of Ecological Environment and Regional Economic Development--A Case Study of Hunchun Area in Amur Tiger and Leopard National Park. *J. North-East For. Univ.* **2020**, *48*, 76–80.
67. Huang, Z.; Chen, X. Livelihood capitals and the migration satisfaction of rural households—A case in southern Shaanxi province. *J. Arid Land Resour. Environ.* **2018**, *32*, 47–52.
68. Huang, Z.; Li, J. Impact of the follow-up supportive policies of relocation for poverty alleviation on multidimensional relative poverty of farming households. *Resour. Sci.* **2022**, *44*, 1905–1917. [CrossRef]
69. Huang, Z.; Chen, X.; Li, J. Research on the impact mechanism of ecological migration policy on rural household income based on the Structural Equation Model with formative indicators. *Resour. Sci.* **2018**, *40*, 439–451.
70. Sander, T.; Lee, T.P.; Univ, L. Smartpls for the Human Resources Field to Evaluate A Model. In Proceedings of the New Challenges of Economic and Business Development, Riga, Latvia, 8–10 May 2014; pp. 346–358.
71. Hassan, M.G.; Othman, A.A.; Ismail, M.A. Effect of Relational-Oriented Exchange on Strategic Performance of Outsourcing Success Using SmartPLS. *Adv. Sci. Lett.* **2015**, *21*, 1131–1134. [CrossRef]
72. Abdullah, N.D.; Malik, S.A. Development of Business Failure Prediction Model Using SmartPLS on Listed Construction Companies on Bursa Malaysia. *Adv. Sci. Lett.* **2018**, *24*, 2599–2603. [CrossRef]

73. Hair, J.F. Reflections on SEM: An Introspective, Idiosyncratic Journey to Composite-Based Structural Equation Modeling. *Data Base Adv. Inf. Syst.* **2021**, *52*, 101–113. [CrossRef]
74. Al-Khateeb, B.A.A.; Dahalin, Z.B.M. Re-Investigating the Relationship between Information Source, Information Channels and Information Choice Strategies Using the SmartPLS. In Proceedings of the Creating Global Competitive Economies: 2020 Vision Planning & Implementation, Rome, Italy, 13–14 November 2013; Volume 1–3, pp. 1797–1805.
75. Bujang, M.A.; Omar, E.D.; Baharum, N.A. A Review on Sample Size Determination for Cronbach's Alpha Test: A Simple Guide for Researchers. *Malays. J. Med. Sci.* **2018**, *25*, 85–99. [CrossRef] [PubMed]
76. Zhang, L.Y.; Jiang, W.J.; Chen, G. A Note on Bartlett's M Test for Homogeneity of Variances. *Commun. Stat. Theory Methods* **2010**, *39*, 2890–2900. [CrossRef]
77. Charway, H.; Bailer, A.J. Testing multiple-group variance equality with randomization procedures. *J. Stat. Comput. Simul.* **2007**, *77*, 797–803. [CrossRef]
78. Hossain, M.I.; Jamadar, Y.; Islam, M.F.; Rashed, M.; Akter, T. Environmental sustainability practices in SMEs: Insights from integrated PLS-SEM and fsQCA approaches. *J. Clean. Prod.* **2025**, *503*, 145185. [CrossRef]
79. Ma, Z.; Wei, X.; Chen, W. Study on the effect of wildlife damage compensation on farmers' willingness to sustain cultivation under the dual objective constraint of conservation and development. *J. Nat. Conserv.* **2025**, *86*, 126938. [CrossRef]
80. Yu, L.; Shi, H.; Ren, S.; Zhang, Q. Farmers' adaptive behaviors towards wildlife incidents regarding national park management: The role of informal institutions. *J. Rural Stud.* **2024**, *111*, 103418. [CrossRef]
81. Li, F. Conflicts of Interest in National Parks in China: Category Explanation, Generation Mechanism, and Relief Path. *Nat. Prot. Areas* **2023**, *3*, 11–20.
82. Chen, Z.; Cheng, Y.; Zhou, M.; Xu, L.; Qi, X. Local residents' cognition, willingness and influencing factors on wildlife conflict in national parks: A case study of Wuyishan National Park, China. *Acta Ecol. Sin.* **2022**, *42*, 2698–2707. [CrossRef]
83. Xiao, L.; Liu, Q.; Yu, H.; Lin, M. Community regulation in national park based on land use conflict identification:A case study on Qianjiangyuan National Park. *Acta Ecol. Sin.* **2020**, *40*, 7277–7286.
84. Wianti, K.F. Land Tenure Conflict in The Middle of Africa van Java (Baluran National Park). In Proceedings of the 4th International Conference on Sustainable Future for Human Security Sustain 2013, Kyoto, Japan, 19–21 October 2013; pp. 459–467.
85. Penteriani, V. Conflict animals or conflict people-that is the question. *Hum.-Wildl. Interact.* **2023**, *17*, 134–137.
86. Yasmi, Y.; Colfer, C.J.P.; Yuliani, L.; Indriatmoko, Y.; Heri, V. Conflict management approaches under unclear boundaries of the commons: Experiences from Danau Sentarum National Park, Indonesia. *Int. For. Rev.* **2007**, *9*, 597–609. [CrossRef]
87. Depraz, S.; Laslaz, L. Conflicts, acceptance problems and participative policies in the national parks of the French Alps. *Eco Mont-J. Prot. Mt. Areas Res.* **2017**, *9*, 46–56. [CrossRef]
88. Saito, C.H.; Brasileiro, L.; Almeida, L.E.d.; Tavares, M.C.H. Conflitos entre macacos-prego e visitantes no Parque Nacional de Brasília: Possíveis soluções. *Soc. Nat.* **2010**, *22*, 515–524. [CrossRef]
89. Ma, B.; Yan, D.; Wen, Y. Farmers' willingness and influencing factors on purchasing wildlife accident insurance in national parks: A case study of Giant Panda National Park, China. *Acta Ecol. Sin.* **2023**, *43*, 4202–4211. [CrossRef]
90. Zhu, D.; Zhong, L.; Yu, H. Research progress of community development of national parks and implications. *Resour. Sci.* **2021**, *43*, 1903–1917. [CrossRef]
91. Hou, S.; Liao, L.; Liu, K.; Teng, L.; Shen, S. Analysis of community livelihood capital differences and the influencing factors in the Wuyishan National Park. *Nat. Prot. Areas* **2024**, *4*, 20–35.
92. Li, J.; Tang, H.; Kuang, F. Exploring Livelihood Strategies of Farmers and Herders and Their Human Well-Being in Qilian Mountain National Park, China. *Sustainability* **2023**, *15*, 8865. [CrossRef]
93. Qin, Z.; Zhang, Y.; Du, S. Research on Sustainable Livelihoods of Residents in National Park Community: A Case Study on Bawangling Area in the National Park of Hainan Tropical Rainforest. *Nat. Prot. Areas* **2023**, *3*, 1–14.
94. Liu, H. Research on Optimization of Human-land Relationship in National Park Communities Based on Order Parameters: A Sustainable Livelihood Perspective. *Chin. Landsc. Archit.* **2022**, *38*, 89–92.
95. Zhong, J.; Ma, Y.; Mi, W.; Shi, J.; Jiang, X. Livelihood Vulnerability of Herdsmen in Qinghai Lake National Park under Multiple Pressures. *Econ. Geogr.* **2024**, *44*, 169–177.

land

MDPI

Article

Digital Geospatial Twinning for Revaluation of a Waterfront Urban Park Design (Case Study: Burgas City, Bulgaria)

Stelian Dimitrov [1], Bilyana Borisova [2], Antoaneta Ivanova [1], Martin Iliev [1], Lidiya Semerdzhieva [1,*], Maya Ruseva [3] and Zoya Stoyanova [4]

1 Geospatial Systems and Technologies Department, Sofia University St. Kliment Ohridski, 1164 Sofia, Bulgaria; stelian@gea.uni-sofia.bg (S.D.); antoaneti1@uni-sofia.bg (A.I.); martin@gea.uni-sofia.bg (M.I.)

2 Landscape Ecology and Environmental Protection Department, Sofia University St. Kliment Ohridski, 1164 Sofia, Bulgaria; billiana@gea.uni-sofia.bg

3 "Strategic Development" Directorate, 26 Alexandrovska Str., 8000 Burgas, Bulgaria; m.ruseva@burgas.bg

4 "Strategic Planning, Digitalization and Sustainable Urban Development" Department, 26 Alexandrovska Str., 8000 Burgas, Bulgaria; z.stoyanova@burgas.bg

* Correspondence: l.nikolaeva@gea.uni-sofia.bg

Abstract: Digital twins play a crucial role in linking data with practical solutions. They convert raw measurements into actionable insights, enabling spatial planning that addresses environmental challenges and meets the needs of local communities. This paper presents the development of a digital geospatial twin for a residential district in Burgas, the largest port city on Bulgaria's southern Black Sea coast. The aim is to provide up-to-date geospatial data quickly and efficiently, and to merge available data into a single, accurate model. This model is used to test three scenarios for revitalizing coastal functions and improving a waterfront urban park in collaboration with stakeholders. The methodology combines aerial photogrammetry, ground-based mobile laser scanning (MLS), and airborne laser scanning (ALS), allowing for robust 3D modeling and terrain reconstruction across different land cover conditions. The current topography, areas at risk from geological hazards, and the vegetation structure with detailed attribute data for each tree are analyzed. These data are used to evaluate the strengths and limitations of the site concerning the desired functionality of the waterfront, considering urban priorities, community needs, and the necessity of addressing contemporary climate challenges. The carbon storage potential under various development scenarios is assessed. Through effective visualization and communication with residents and professional stakeholders, collaborative development processes have been facilitated through a series of workshops focused on coastal transformation. The results aim to support the design of climate-neutral urban solutions that mitigate natural risks without compromising the area's essential functions, such as residential living and recreation.

Keywords: digital twin; spatial analysis; urban park; natural hazards; carbon sequestration; urban planning; stakeholders; Black Sea coast; geodesign

1. Introduction

The challenges of the Anthropocene in urban management are sparking scientific discussions on creating new criteria for spatial and urban development, developing new methods for data collection, and analyzing the areas occupied by cities. They promote interdisciplinary collaboration to enhance the effectiveness of urban planning decisions through quick yet sustainable solutions that align with the local characteristics of the environment and the region's natural functions [1]. Respect for historically formed knowledge [2], supported by adequate and comprehensive spatial analysis [3], has the potential to refine our spatial urban planning decisions and contribute to the development of urban vitality [4]. The latter focuses on urban and peri-urban forests, which play an increasingly important role in balancing the urban structure [5] and providing valuable ecosystem services for maintaining urban ecosystems [6,7], mitigating the adverse effects of the urban climate [8], improving our living environment, and the quality of green-blue infrastructures [9].

Urban parks are defined as designated open space areas, mainly consisting of vegetation and water, and typically reserved for public use [10]. Urban parks are a significant structural element of urban green infrastructure and are considered to have the highest impact on society, human health and well-being, air and water quality, maintenance of the biodiversity inherent to the area, and, last but not least, social cohesion [10,11].

Of particular interest is managing the benefits of coastal city parks. These parks are highly vulnerable to natural and human-made changes, as they lie at the frontline of a combination of geological (coastal erosion, landslides, soil swelling, and liquefaction) and ecological risks in response to increasing threats from climate change [12], including heat waves, storms, floods, and the spread of invasive species [13]. At the same time, they play a crucial role in attracting and retaining the population [14]. The functionality of waterfront cities, in the broad sense of coastal cities, has historically changed [15] under the influence of geopolitical factors, trade and transport relations, and the industry based on them, or priorities in the service sector, most often related to tourism. This development is accompanied by strong pressure on available natural resources, which, against the backdrop of natural and anthropogenic risks, brings to the forefront of public attention the need for coastal zone regeneration, with a particular focus on restoring and improving urban ecosystem services [16]. However, the success of strategies and plans for adapting coastal cities to climate change depends on a delicate balance with other vulnerabilities in spatial urban development—socioeconomic sensitivity, infrastructure provision, and overall adaptive capacity [17]. Published knowledge on specific types of adaptation in coastal cities, such as ecosystem adaptation, is still limited [12].

Urban coastal green spaces are defined as an emblematic element of the urban landscape, and their functionality in a complex of public benefits (public health, protection of natural and cultural heritage, water and air quality, natural risk regulation) depends on the structural composition and configuration of landscape elements [18] and their integration into the overall green system of the city [16,18]. There is growing attention to the potential of urban forests as nature-based solutions, but delivering effective, practical results depends on addressing a series of challenges, including the provision of region-specific information and the need for interdisciplinary collaboration [19]. Properly addressing contextual sensitivity and taking an individual approach to planning urban forests as nature-based solutions requires synthesizing current information at a scale suitable for the project. This includes considering topographical features, hydrometeorological conditions, soil and biogeographical patterns, the current functionality of the area, ecological risks, and the continuity and heritage of the cultural landscape. High expectations for urban forests are justified when their planning, construction, and maintenance are carried out collaboratively, especially between local authorities and communities [19]. This involves considering land use and competing interests, engaging the public, and exploring opportunities for long-term financing with clear guarantees from relevant local authorities or economic actors involved in the territory's use.

We see such an opportunity in the potential of modern geospatial methods and technologies for providing up-to-date and accurate data on the urban mosaic and infrastructure functionality, which allow for a better understanding of the changing geography of cities, proactive response to environmental risk, to the changing needs of the community, and the development of effective solutions in urban planning in a delicate compromise between the available priorities in urban development [20]. There is growing attention and research focus on the application of digital twins in the field of spatial urban planning and design to optimize the information provision process, including the collection, processing, analyzing, and presenting the flow of data and information from the real world, thus serving as a tool to support a climate-neutral and sustainable city [21].

Combining digital twin (DT) technologies with Geographic Information Systems (GIS) represents a significant advancement in geospatial analysis. This integration allows for continuous synchronization, resulting in a precise depiction of the physical environment.

Continuous synchronization is essential for maintaining a precise and up-to-date representation of the physical environment. While more frequent synchronization demands significant investments in terms of time, human resources, and computational power, it ensures high efficiency and prevents errors associated with outdated information. Digital twins that are not adequately updated lead to inaccurate analyses, undermine informed decision-making, and carry the risk of implementing plans that do not correspond to current realities, thereby resulting in significant financial and operational challenges. For instance, planning new tree planting based on a digital twin that fails to reflect recent tree removals or new infrastructure would lead to wasted resources and incorrect assessments of green space. It provides decision-makers with dynamic and context-specific insights for more informed decision-making [22].

While this study focuses on spatial urban planning applications, the digital geospatial twin methodology demonstrates significant potential across diverse disciplines and sectors. Beyond urban planning, these technologies support infrastructure management through predictive maintenance, environmental monitoring for conservation efforts, and energy sector applications, including renewable energy planning. Additional applications span emergency management for disaster preparedness, heritage preservation for archaeological documentation, and public health analysis for community wellbeing assessment. This methodological framework's adaptability stems from its core strength of transforming complex spatial data into actionable insights for evidence-based decision-making across any field requiring precise environmental understanding.

Geospatial data are crucial for the design of green urban areas in reflection of local environmental conditions [23]. Digital geospatial twins can facilitate the development of a comprehensive process of interaction with stakeholders in the planning of green infrastructure and the development of intervention scenarios in response to environmental challenges and the compromise-based development of solutions for urban sustainability [24]. This type of spatial planning, which uses geospatial data and analyzes them through geospatial technologies with the active participation of stakeholders, is rooted in the geodesign framework. Data-driven participation methods, such as geodesign, are very promising for supporting strategic planning to make cities and regions more sustainable [25]. Social learning and collective action, as well as geodesign approaches that apply systems thinking using geographical knowledge, are considered key factors for urban transformation that can better provide the qualities valued and needed by society [26]. There are several examples of geodesign practices in resilience planning and resilience thinking with a focus on climate change, disaster risk reduction, and management activities such as floods and sea level rise risks [25,27].

This scientific study is conducted to support the city of Burgas, the main port city on the southern Bulgarian Black Sea coast, in designing climate-neutral urban planning solutions for the improvement and revitalization of the coastal area (at the Sarafovo residential district). The study aims to provide up-to-date data on the territory by developing a digital geospatial twin and testing the application of the newly obtained data in the development of scenarios for the redesign and renovation of the coastal zone in active interaction with stakeholders.

The following tasks were set during the study: (1) Identification of the current topography of the terrain and vegetation structure to discuss the advantages and limitations concerning the desired functionality of the coastline in response to urban priorities, public needs, and providing an adequate response to contemporary challenges; (2) Monitoring the condition of sites affected by geological risk for targeted planning and design for long-term management; (3) Supporting the calculation of the carbon footprint in scenarios for coastal vegetation restoration; (4) Appropriate visualization of the results obtained to encourage communication with stakeholders.

2. Materials and Methods

The study is based on the implementation of a comprehensive spatial model, developed and applied based on the digital geospatial twin approach, which was created for the Sarafovo district of Burgas and a consistent thematic processing of the results for the development of well-founded data-driven solutions for the improvement of the territory revitalization of the coastal functions. The study follows the principles of the geodesign framework [28–30], and based on this, the digital geospatial twin here acts as an geospatial information and simulation design hub [31], facilitating interactions between geographers, landscape architects and ecologists, urbanists, urban planners, municipal administration, architects, local businesses, and environmental and cultural non-governmental organizations in co-developing scenarios based on climate-neutral urban design principles.

2.1. Site Properties and Conditions

Burgas is the fourth largest city in Bulgaria (253.6 km^2) with a population of 188,242 people (average population density—742 people/km^2) [32]. It is located on the southern Black Sea coast of the country and is of major importance to Bulgaria's agricultural sector, chemical industry (the largest oil refinery in Southeast Europe), transport and logistics, trade, and tourism. The territory falls within the Black Sea climatic region of the country, with an average annual temperature of 12.7 °C and average annual precipitation of 553.7 mm, with the maximum in November. In spatial terms, despite the predominantly flat nature of the relief, the territory of Burgas city is extremely complex, with a high degree of fragmentation of its parts, mainly due to the density of extensive water bodies—the Atanasovsko, Vaya, and Mandrensko lakes, as well as the Black Sea waters. This geography has determined the development and configuration of the settlement network—seven residential complexes and two new ones under development, as well as eleven quarters, among which is the Sarafovo quarter (Figure 1).

The vision of Burgas for spatial development emphasizes preserving local identity while ensuring access to modern, resource-efficient, climate-adaptive, and competitive economic opportunities. This involves investing in the enhancement of urban spaces, connectivity, and the preservation of a quality natural environment and green infrastructure. The basis for the implementation of these complex tasks is the creation of conditions for increasing the connectivity of green-blue infrastructure by utilizing the advantages of the coastal strip (Compact Zone for Integrated Intervention "Coastal Zone") [33].

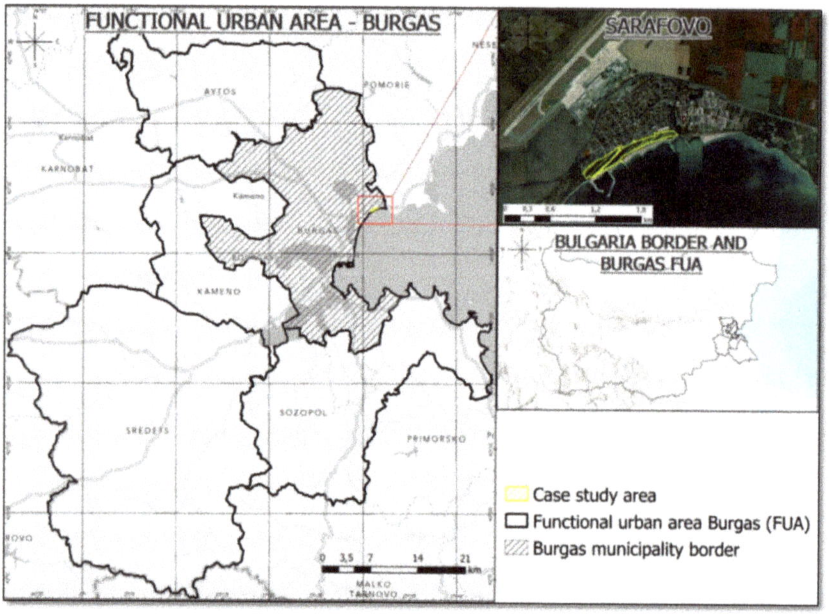

Figure 1. Function Urban Area—Burgas city.

The Sarafovo neighborhood (3500 permanent residents) is located in the northeastern part of Burgas, on the Black Sea coast, and is geographically relatively isolated from the rest of Burgas, even though it is only 10 km away from the city center. To the north of it runs the main Republic Road I-9, and to the northwest is Burgas Airport for passenger transport, both of which create spatial constraints on territorial expansion. In the southwestern and southeastern directions, there are areas designated by the General Development Plan for Residential Construction, which are gradually being implemented and are expected to undergo intensive development soon. The selected project area is located parallel to the coastline, south of the Sarafovo district. The total area is 140,974.8 m^2, divided into two properties. It was selected based on the strategic goal of achieving unified connectivity of green areas along the coastline of Burgas, the desire to build perpendicular connections to the interior, identified needs, and opportunities for intervention in the near future. The first step in this direction is the improvement of the coastal zone, which today has partially degraded vegetation and disturbed terrain (designated as "Public Settlemental Park, Garden"). According to the Master Plan, the coastal land plots are designated for landscaping, which provides a good opportunity for development.

The geographical location and natural features of Burgas Municipality, the mild climate, and extensive water bodies favor the presence of rich biological diversity. The case-study area does not fall within the boundaries of Protected Zones and NATURA 2000 Protected Sites, but it is in spatial proximity to them and can serve a supporting function. To the west lies the protected area of Atanasovsko Lake, a key location on the Via Pontica migratory route for birds in Europe (the northern part of the lake is a maintained reserve (IV, IUCN), while the southern part is the protected area of Burgas Saltworks (VI, IUCN). It is an internationally significant wetland under the Ramsar Convention and BirdLife International. To the east is the Pomorie Lake protected area, also of exceptional value in terms of biodiversity (Figure 2).

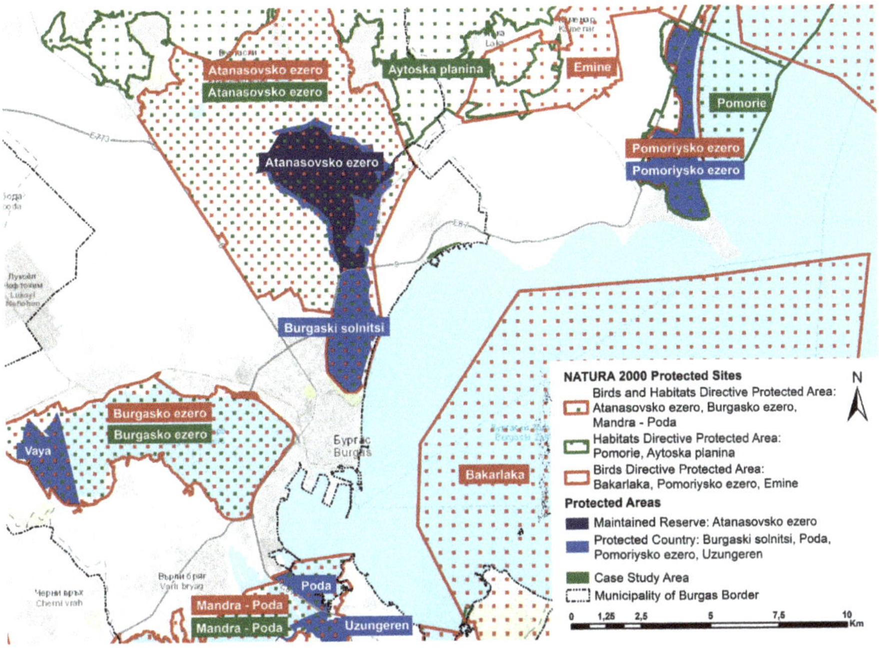

Figure 2. Protected Areas, NATURA 2000 Protected Sites.

Several key challenges have been identified within the area of Sarafovo Quarter, including high density and intensity of development, hindering the connection of coastal greenery with that in and outside urbanized areas, a high proportion of seasonally occupied dwellings, uneven distribution of green areas for public use (such as parks and gardens), and insufficient space relative to the population—2.4 m^2/person with a minimum norm of 20 m^2/person. The quarter faces issues related to negative impact on biodiversity, noise and dust pollution (all from the proximity to the airport and major road and building characteristics), as well as geological risk in the conditions of a steep coast (130 m from the coast, on land the height increases to 35 m)—coastal erosion, landslides, soil swelling, and, in the immediate vicinity, liquefaction of weak soils (Figure 3). The latter creates a direct need for surface and groundwater management to preserve and strengthen relief forms.

The pilot territory can be defined as undeveloped—there is no alley network and technical infrastructure. In the past decades, numerous anthropogenic interventions have been made in the area, such as land filling from construction, deforestation, artificial leveling of the terrain, parking, and the creation of illegal access paths.

The technical project Coastal Park near Sarafovo quarter was developed in 2009 and included the construction of reinforcement systems, a network of alleys, and a wide range of themed spaces. The project was not implemented due to a low degree of respect for the topography and the need for serious interventions in the terrain, and at present, the project is not applicable as a result of the serious natural and anthropogenic changes in the landscape. The redevelopment of the conceptual and technical design requires up-to-date and detailed spatial data, carried out in consultation with a wide range of stakeholders, including representatives of key professions, educational and scientific institutions, administrative bodies, local communities, non-governmental organizations, and businesses (beach concessionaires, salt producers, and active parties involved in the operation of Sarafovo Fishing Port).

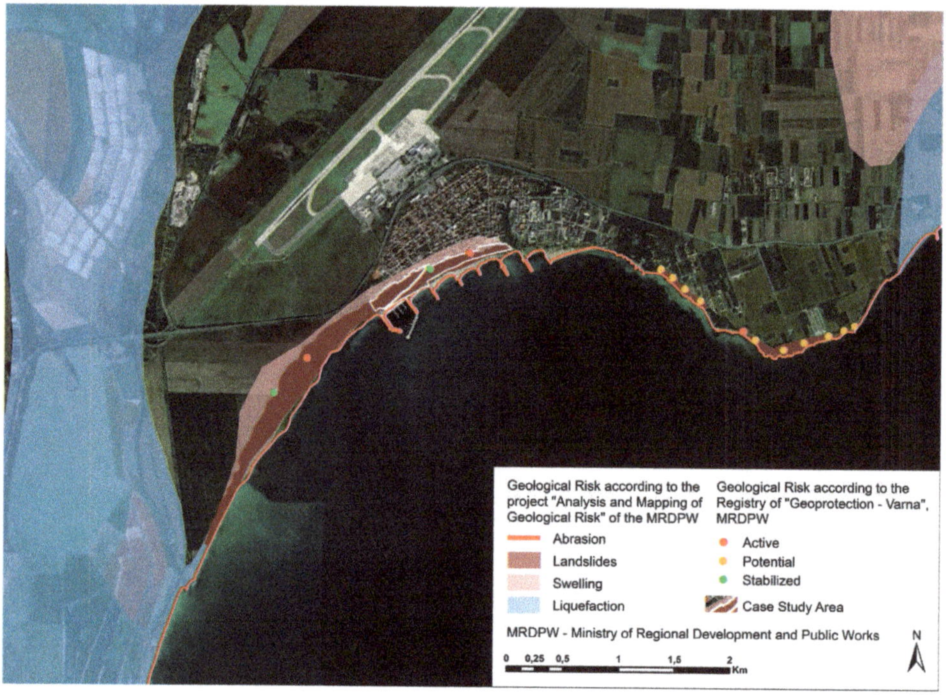

Figure 3. Geological risk of the territory.

2.2. *Digital Geospatial Twin—Field Observations*

The creation of a digital twin for the Sarafovo district is grounded in a multi-tiered geospatial data acquisition approach. The primary objective of the field campaign was to generate accurate, high-resolution spatial datasets for characterizing both built and natural environments. The adopted methodology integrates aerial photogrammetry, ground-based mobile laser scanning (MLS), and airborne laser scanning (ALS), enabling robust 3D modeling and terrain reconstruction under varying land cover conditions (Table 1). Each method was selected to address specific observational challenges: while photogrammetry excels in capturing visual and structural detail in open areas, it struggles in densely vegetated zones; conversely, LiDAR systems provide elevation and volumetric data even in occluded or shadowed areas. The combination of these complementary methods supports a more holistic and precise representation of the study area. All spatial datasets were referenced to the Bulgaria Geodetic System 2005—BGS2005 coordinate system (EPSG:7801).

Table 1. Summary of data acquisition techniques for the digital twin development.

Feature	Aerial Photogrammetry	Ground-Based Mobile Laser Scanning	Airborne Laser Scanning (ALS)
Method	Structure-from-Motion (SfM) Photogrammetry	Mobile Laser Scanning (MLS) with SLAM	Airborne Laser Scanning (LiDAR)
Platform	SenseFly eBeeX (Fixed-wing UAS, AgEagle Aerial Systems Inc., Wichita, KS, USA)	Operator on foot and bicycle	Multirotor UAS
Sensor	SenseFly AeriaX (High-resolution RGB camera, AgEagle Aerial Systems Inc., Wichita, KS, USA)	GeoSLAM ZEB Horizon (LiDAR with integrated RGB camera, GeoSLAM, Nottingham, UK)	mdLiDAR1000HR (LiDAR, Microdrones, Madison, AL, USA)

Table 1. *Cont.*

Feature	Aerial Photogrammetry	Ground-Based Mobile Laser Scanning	Airborne Laser Scanning (ALS)
Primary Data	1223 high-resolution nadir RGB images with precise geotags.	Georeferenced 3D point cloud and synchronized RGB imagery.	High-density 3D point cloud
Key Deliverables	-Orthophoto Map (2.5 cm) -Dense RGB Point Cloud (~208 M points) -Digital Surface Model (DSM) -Digital Terrain Model (DTM)	-Colorized Point Cloud (~53 M points) -High-accuracy DTM (sub-5 cm) -Topographic cross-sections and profiles	-Classified Point Cloud (multiple returns) -High-accuracy DTM (10 cm)
Advantages	-Excellent for capturing realistic color and texture. -Highly efficient for mapping large, open areas. -Produces high-resolution, visually intuitive orthophotos. -Cost-effective for generating initial base layers.	-Superior at capturing data under dense vegetation. -Excellent for modeling vertical surfaces (facades) -Provides very high detail and accuracy at ground level. -Flexible and portable in complex/confined spaces.	-Penetrates vegetation to accurately map the bare-earth. -Provides highly accurate elevation data (Z-values). -Unaffected by shadows or ambient light conditions. -Efficiently fills data gaps over large, inaccessible areas.
Disadvantages	-Struggles to penetrate dense vegetation canopy. -Data quality is affected by shadows and poor light. -Poor at capturing vertical surfaces from nadir flights. -Can produce geometric distortions in occluded areas.	-Limited spatial coverage can be time-consuming. -Inaccessible in very overgrown terrain. -SLAM accuracy can drift in large, featureless areas. -Data can be noisy due to moving objects (people).	-Lower point density on vertical surfaces. -Canopy data quality can be affected by wind. -Higher equipment and operational cost.

2.3. *Aerial Photogrammetric Mapping*

An aerial photogrammetric survey was conducted in August 2022 over the Sarafovo area using the Ag Eagle eBeeX (AgEagle Aerial Systems Inc., Wichita, KS, USA) (Figure 4a). This fixed-wing unmanned aerial system (UAS) was strategically selected over a multirotor alternative due to its superior flight endurance and efficiency, making it ideal for mapping large, contiguous areas like the entire Sarafovo quarter with minimal operational interruptions. The platform was equipped with the AeriaX photogrammetric camera, chosen for its high image fidelity and minimal geometric distortion. Flight planning and mission execution were managed using Ag Eagle's official flight management software, eMotion (version 3.20). The flight plan includes a horizontal mapping mission configured for 80% frontal and 70% lateral image overlaps at a nominal altitude of 120 m above ground level (AGL), achieving a ground sampling distance (GSD) of 3 cm/pixel.

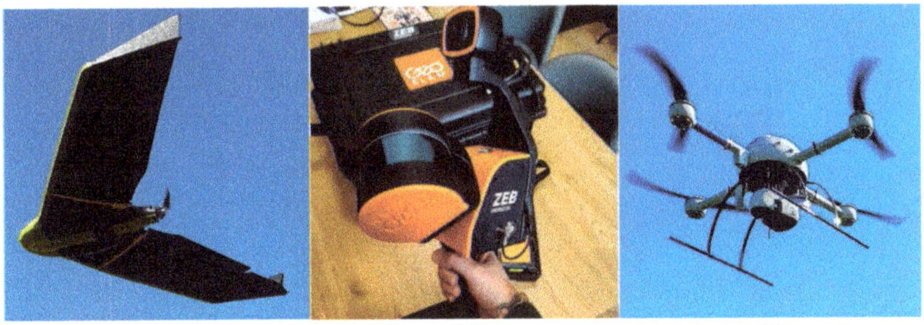

(**a**) (**b**) (**c**)

Figure 4. Remote Sensing Equipment: (**a**) Ag Eagle eBeeX Fixed-Wing UAS; (**b**) GeoSLAM Zeb Horizon LiDAR; (**c**) mdLiDAR1000HR Multirotor UAS.

The mapping was conducted on a sunny day with minimal wind, as these conditions are critical for ensuring consistent illumination and maintaining the stability of the UAS platform. A total of 1223 nadir aerial images were captured under clear sky conditions to reduce shadowing and radiometric variation. Image processing was carried out using Pix4Dmapper (version: 4.5.6). This software leverages Structure-from-Motion (SfM) algorithms, a computational method chosen for its proven ability to derive high-density 3D data from unstructured 2D image sets. The SfM process analyzes the overlapping images to simultaneously determine the camera's position for each shot and reconstruct the scene's geometry, which is the foundational step for generating the dense point cloud and subsequent products.

The processing resulted in a comprehensive suite of geospatial deliverables. The foundational dataset is a dense, RGB-classified 3D point cloud containing approximately 208 million points, which provides a detailed 3D representation of all surface features. From this point cloud, two key elevation models were generated: a 10 cm resolution Digital Surface Model (DSM) capturing the top-most surfaces of buildings and vegetation, and a corresponding Digital Terrain Model (DTM) representing the bare-earth topography after filtering out non-terrain objects. Finally, a seamless, true-color orthophoto map was produced at 2.5 cm resolution, offering a geometrically corrected and distortion-free visual base layer for the study area.

Despite favorable weather, some urban areas displayed reduced accuracy due to shadows cast by high-rise structures and occlusion in narrow streets, leading to localized DSM distortions. These limitations were mitigated through manual editing and data fusion with LiDAR products. RTK GNSS corrections were applied during flight, achieving horizontal accuracy of approximately 2 cm and vertical accuracy of 3 cm. A set of independently measured checkpoints was used for geometric validation. To ensure broad interoperability, these deliverables were exported in industry-standard formats, such as GeoTIFF for the raster models and orthophoto and LAS for the 3D point cloud, allowing for easy use in virtually any geospatial software.

2.4. Ground-Based Mobile Laser Scanning

To complement aerial observations and capture terrain features under vegetation, a ground-based mobile laser scanning (MLS) survey was carried out in March 2024 using the GeoSLAM Zeb Horizon system (Figure 4b). This handheld, SLAM-enabled scanner was selected for its portability, high point density, and suitability in complex, vegetated environments. The system is co-equipped with a synchronized camera, enabling the capture of RGB imagery to colorize the resulting 3D point cloud and enhance visual interpretation. Furthermore, its extended battery life supports continuous data acquisition, which was essential for covering the large and fragmented project area in a single or double field session.

Crucially, its ground-level perspective is also essential for capturing detailed data of vertical surfaces, such as building facades and all vertically positioned infrastructure elements. These features are inherently hidden or poorly represented in nadir-oriented aerial datasets. To ensure comprehensive coverage across the study area's diverse terrain, the survey was conducted using a flexible traversal strategy. The operator proceeded on foot in complex or confined spaces, such as forested zones and narrow pathways, to maximize data resolution and capture fine-scale ground features. A bicycle was employed to efficiently scan long, linear infrastructure like roadways, significantly increasing the speed of data acquisition without compromising quality.

Data acquisition was conducted under stable daylight conditions and with minimal wind to reduce motion artifacts caused by vegetation movement. Furthermore, the survey was intentionally timed to periods of low public activity to avoid noise and "ghosting" artifacts in the point cloud, which are commonly caused by moving pedestrians. The raw point cloud was processed using GeoSLAM Hub software (version: 6.2.1), where SLAM algorithms corrected for drift and optimized spatial alignment.

The primary deliverable from the MLS survey is a high-density, georeferenced 3D point cloud (LAS format) comprising approximately 53 million points. This point cloud is fully colorized using integrated camera imagery and classified to distinguish between ground returns, vegetation, and man-made structures, achieving a vertical accuracy of sub-5 cm. Based on the filtered ground points, a 10 cm-resolution Digital Terrain Model (DTM) was generated, providing a detailed representation of the bare-earth topography. For direct engineering use, specific topographic cross-sections and profiles were also extracted as vector datasets to support design applications.

The system successfully captured fine-scale topography and ground features beneath dense vegetation, but terrain inaccessibility in several overgrown zones limited scanning coverage. Manual segmentation and noise filtering were employed during post-processing to resolve ambiguities caused by vegetation clusters and underbrush.

2.5. *Airborne Laser Scanning*

An airborne laser scanning (ALS) survey was conducted in June 2024 to enhance vertical accuracy and resolve remaining spatial data gaps, especially in areas where photogrammetric and ground-based methods were constrained. The mission used a mdLiDAR1000HR system mounted on a multirotor unmanned aerial system (UAS) (Figure 4c). This platform was strategically chosen for its ability to fly slowly and at low altitudes over complex terrain, which is essential for achieving high point density, while the LiDAR sensor's capacity to penetrate vegetation canopy makes it superior to photogrammetry in forested zones. The system integrates high-frequency pulse-based LiDAR with precise GNSS/IMU positioning (RTK/PPK).

A pre-defined grid flight pattern with 80% sidelap was executed at 100 m AGL with a 70° field of view, producing a point density of 150–300 points/m^2 to ensure complete area coverage. The flights were performed under calm and clear weather conditions to minimize UAS drift and wind-induced vegetation motion. For safety reasons and due to the operational weight of the platform, the survey area was secured and cleared of all unauthorized personnel during the mission. This focused particularly on vegetated slopes, infrastructure corridors, and transitions between open and forested terrain.

The post-processing of the raw LiDAR data yielded a set of key deliverables. The primary output is a classified 3D point cloud (LAS/LAZ format), which includes multiple returns (first, intermediate, last) essential for detailed canopy and ground characterization. From this, a 10 cm resolution Digital Terrain Model (DTM) was derived using progressive morphological ground filtering algorithms.

Although the ALS system achieved consistent performance, wind-induced vegetation motion introduced noise into the canopy point cloud, and terrain steepness in certain segments affected return density uniformity. These were addressed through trajectory adjustment, filtering, and multi-pass flight data merging.

2.6. *Using Digital Geospatial Twin Results to Calculate Carbon Footprint in Coastal Development and Improvement Scenarios*

Our previous studies demonstrate the vulnerability of the city of Burgas to climate change, particularly regarding its exposure to the urban heat island effect [34]. Utilizing the geographical characteristics of the Sarafovo district, we analyzed the area, structure, and condition of contemporary vegetation using data from a digital geospatial twin. This analysis enabled us to calculate the potential contribution of vegetation to carbon sequestration under various coastal development scenarios.

For the study, the "Methodological guidelines for the preparation of consolidated documentation to prove climate resilience (including climate change mitigation and adaptation)" of project proposals under the Operational Program "Environment 2021–2027" of the Republic of Bulgaria [35] were applied. The urban coastal park project discussed here aligns with the concept of "Green Infrastructure in Urban Environments." Its sustainability is evaluated based on the carbon absorption capabilities of growing biomass. The planned green spaces will feature a variety of trees, shrubs, and grasses, as well as combinations of these elements. The absolute emissions of the project are calculated for a standard year of operation: t CO_2e/yr = average annual carbon sequestration (t CO_2e/yr).

Average annual carbon sequestration by tree species is calculated using the following formula:

$$Average\ annual\quad carbon\ sequestration\left(\frac{tCO^2e}{year}\right) = \left[MAI\left(\frac{\frac{m^2}{ha}}{year}\right)\right] \times [BCEF] \times [1+R] \times \left[CF\left(\frac{t\ C}{t\ dry\ matter}\right)\right] \times \left[CCF\left(\frac{tCO^2e}{t\ C}\right)\right] \times [Forest\ area\ (ha)]$$

where the following sources of information are applied:

MAI (m^3/ha/year) (mean annual increment). The data presented here are taken from the IPCC Guidelines for National Greenhouse Gas Inventories, Chapter 4—Forests [36], using the average value for hardwoods in temperate climates. The resulting MAI value is 0.956 m^3 (for deciduous species) and 0.706 m^3 (for coniferous species).
BCEF (biomass conversion and expansion factor). BCEF values are consistent with EIB Project Carbon Footprint Methodologies, 2023 [37]: 0.621 t/m^3 (for deciduous species) and 0.464 t/m^3 (for coniferous species).
R (below-ground biomass to above-ground biomass ratio). R is estimated conservatively based on expert knowledge and is consistent with the EIB Project Carbon Footprint Methodologies 2023 [37]: 0.24 (for deciduous species) and 0.29 (for coniferous species).
CF (tC/t dry matter) (carbon fraction). The value used here is in line with the EIB Project Carbon Footprint Methodologies, 2023 [37]: 0.48 (for deciduous species) and 0.51 (for coniferous species) tC/t dry matter.
CCF (tCO_2e/tC) (carbon conversion factor)—The value used here is in accordance with the Methodological Guidelines: conversion factor for C to tCO_2e = (12 + (16 × 2))/12 = 3.67. CCF = 3.67 tCO_2e/tC
Forest area (ha)—Different values assumed in the development scenarios are used here. For the zero scenario, the carbon footprint of existing vegetation was calculated as follows: broadleaf tree species cover 69,550 m^2, coniferous tree species cover 4968 m^2, and the current area covered by shrubs and grasses is 24,839 m^2. For scenario 2, the calculations were made using the following values: broadleaf tree species 3740 m^2, coniferous tree species 910 m^2, and 12,092 m^2, according to the planned area for landscaping with shrub and grass species.

The assessment of emissions/removals from shrub and grass species is based on an assessment of changes in carbon stocks in living biomass and soil. The approach is consistent with the National Greenhouse Gas Inventory [38], where it is assumed that all changes in biomass carbon stocks occur during the first year. To calculate changes in annual carbon stocks in living biomass, the following equation is applied:

$$The\ annual\ change\ of\ carbon\ stock\ in\ biomass = A_{convers}\left(L_{conversion} + \Delta C_{growth}\right)$$

where the following sources of information are applied:

$$L_{conversion} = \mathrm{C}_{\mathrm{after}} - \mathrm{C}_{\mathrm{before}}$$

$$L_{conversion} = C_{after} - C_{before}$$

$$A_{convers} - annual\ areas\ of\ the\ lands\ converted\ to\ grassland,\ ha\ yr^{-1}$$

$$L_{conversion} - carbon\ stock\ in\ the\ biomass\ of\ lands\ which\ were\ converted\ to\ grassland,\ tonnes\ C\ ha^{-1}$$

ΔC_{growth}
$-$ *change of the carbon stock in the biomass in the first year after the conversion tonnes C* ha^{-1}
$-$ 5.24 tC/ha for shrubs and grasses

$$C_{after} = 0$$

$$C_{before} = 3.56\ \mathrm{tC/ha\ for\ anuual\ crops\ (calculated\ for\ Bulgaria)}$$

Soil carbon stocks 20 years after conversion to grassland are assumed to be 86.96 tC/ha, i.e., 4.35 tC/ha/yr.

For a comprehensive assessment of carbon sequestration, the results of the two formulas are summarized here, which corresponds to the guidelines of the methodology [35].

Field analyses have shown that the broadleaf tree vegetation currently consists of Acer campestre, Fraxinus angustifolia, Morus alba, Juglans Regia, Ficus carica, Prunus dulcis, Cydonia oblonga, Pyrus communis, Celtis australis, as well as Robinia pseudoacacia, Gleditsia triacantha, and Maclura pomifera. Among the conifers are Cedrus libani and Pinus nigra. The shrubs found here include Rosa canina, Rubus caesius, Prunus spinosa, Daphne mezereum, Elaeagnus angustifolia, Sambucus nigra, Paliurus spina-christi, Humulus lupulus, and Clematis vitalba. On degraded land, Ailanthus altissima and Amorpha fruticosa are found. The most widely represented grass species are: Agrostis capillaris, Amaranthus albus, Chenopodium ficifolium, Convolvulus arvensis, Xanthium strumarium, Phragmites australis, Cynodon dactylon, Cichorium intybus, and Suaeda maritima.

Scenario 2, which includes plans for new landscaping, aligns with the List of Suitable Tree Species for Parks, Land Improvement, and Forest Plantations established by the Bulgarian Ministry of Environment and Water in 2023. This list is informed by expert opinions regarding the sustainability of urban vegetation, considering the impacts of climate change on Bulgaria's landscape. Scenario 2 features a high percentage of grass areas (72% of the areas designated for landscaping) to strengthen landslide slopes and in locations where terrain terracing is planned. These changes are expected to enhance soil quality, improve drainage, and reduce surface runoff.

2.7. Data-Driven Visualization, Communication, and Co-Creation

To achieve sustainable development despite the potential consequences of ongoing climate change, a sustainability planning methodology is needed that enables and ensures the participation of local communities, policymakers, and other stakeholders. In this regard, geodesign and related methods offer opportunities to improve urban sustainability planning [25]. Co-creation is a key principle, with regular public consultations involving representatives from key professions, educational and scientific institutions, administrative bodies, local communities, non-governmental organizations, and businesses to gather input and collaboratively shape the design and functionality of the waterfront. Data-driven platforms enable transparent decision-making and allow for real-time feedback from the community on environmental conditions and the use of public spaces.

The results of the digital geospatial twin were used in the preparation of meetings and workshops with stakeholders in June 2024 in applicable visualization materials, in the development of routes for field observations, and in the preparation of arguments in discussions of the available limiting factors and opportunities in the intervention area. On 12 June 2024, a series of consultations was held using a web GIS application (ESRI ArcGIS Field Maps, version 24.3.0) (Figure 5). During this event, a field workshop was conducted with a team of experts from various fields involved in the international REVALUE project. Out of a total of 33 participants (12 architects, 7 GIS engineers, 7 municipality representatives, 3 landscape architects, 1 landscape ecologist, 1 ecologist, 1 geographer, 1 urban planner), 12 are university representatives, and had the opportunity to share their ideas for the design and future development of the project area. Each proposal is marked on the ground using the GIS mobile application (ESRI ArcGIS Field Maps, version 24.3.0) and its ability to record the GPS position of the user. Based on a preliminary review of international examples of contemporary waterfront parks, the local administrative team compiled an initial list of thematic zones. In the application, these were presented as a drop-down list (Table 2), and participants in the exercise had the opportunity to add their individual suggestions as notes.

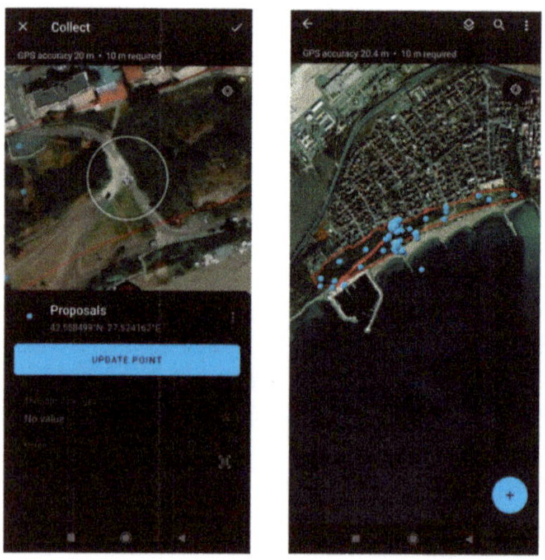

Figure 5. Visualization of the GIS mobile application (ESRI ArcGIS Field Maps, version 24.3.0).

Table 2. Proposals for the development of the project area (for the field workshop).

Type of Proposal	Count
Location identified as important, but no proposal for utilization specified	8
Panoramic terrace	10
Zone for prolonged stays and social interaction	2
Zone for children's playground	2
Zone for sports playground	7
Green arc corridor	3
Zone for afforestation	4
Botanical garden and aroma garden	4
An outdoor cultural events area with terraced seating and stage/small celebrations	1

Table 2. *Cont.*

Type of Proposal	Count
Zone for presenting local life and history—fishing, local fishing communities	1
Zone for local craft products and foods	1
Zone for observing the natural environment, plants, and animals;	1
Zone for ecological education and practice of nature—based solutions	1
Zone for science of geography, geology, climatology. Technological systems for monitoring natural components and biodiversity, including demonstration zones	1
Zone for practice by students in artistic disciplines/Schools, Art centers and National Academy of Art—Burgas	2
Zone for rope construction for children and adults	1
Zone for extreme aerial sports	1
Zone with Monuments	2
Zone for picnic	9
Zone for meditations/with small water feature	4
Zone for off-leash dogs	1
Zone for temporary exhibitions	1
Zone for bicycle rental	4
Zone for parking personal bicycles	5
Zone for parking personal cars	2
Café	2
Other individual proposals outside of the proposed thematic zone list (drinking water fountain, benches, half plant arc for shade, pathways, bike parking station) (Please explain in "notes" field)	13

The 3D models and interactive maps from the Digital Geospatial Twin were applied during public consultations and partner meetings, including the Impact Model Workshop (12 June–14 April 2024), to obtain location-specific needs, desires, and views of stakeholders and experts. Recurring themes highlighted the desire that the park be kept "natural" and to use nature-based solutions to stabilize soils, implement non-invasive, low-impact solutions to improve pedestrian access to the beach, preserve perennial trees and vegetation, remove invasive species, ensure natural air conditioning of the park by promoting marine air corridors, improve viewpoints and limit sealed surfaces, and particularly unregulated parking adjacent to the beach. The intersection of interests is the strengthening of the salt marsh bodies and the drainage of rainwater without affecting tourist activity during the summer season (the Sarafovo district is a popular weekend destination for the residents of the city of Burgas).

3. Results

3.1. Digital Geospatial Twin—New Data on the Territory and Situation Analysis

The developed digital twin of the Sarafovo district represents an integrated, multi-layered model that fuses geospatial data acquired through three distinct methods. It combines the visual information and texture from aerial photogrammetry (orthophoto mosaics and 3D models) with the precise three-dimensional geometry obtained from airborne and ground-based laser scanning (high-density point clouds). This combination of technologies creates a unified digital replica of the physical territory, which accurately maps both the visible elements—such as buildings, infrastructure, and vegetation—and the underlying terrain hidden beneath them. This holistic, multi-dimensional model serves as the foundation upon which the subsequent analyses of the existing conditions were performed.

3.1.1. Land Cover Analysis and Green Infrastructure Deficit

The generated 2.5 cm resolution orthophoto map enabled precise land cover classification, meticulously delineating impervious surfaces (buildings, roads, parking lots) from pervious ones (green spaces, bare soil) (Figure 6). The quantitative analysis of this data, integrated with cadastral and urban planning information, provided empirical evidence of the registered rapid and often inadequately supported urban expansion, linked to the construction of multi-family buildings up to 15 m in height. The most significant finding of this analysis is the objectively established critical deficit of green space. The calculated value of merely 2.36 m^2 per capita falls drastically short of the minimum standard of 20 m^2 per capita, providing an indisputable, data-driven argument for the necessity of targeted interventions to increase green infrastructure.

Figure 6. Surface type classification—based on the processed aerial photogrammetry data.

Data analysis identified ground disturbances, including those caused by illegal vehicle passage and parking, as well as degraded vegetation (Figure 7a,b). The current perspective on the contemporary situation provides a fundamental basis for discussing potential interventions in the coastal zone. These interventions aim to balance the spatial structure of the entire neighborhood, considering its functional characteristics and adjacent features such as the airport, main national road, and the protected Atanasovsko Lake lagoon.

(a) (b)

Figure 7. (**a**) Unauthorized parking areas visualized by orthophoto map; (**b**) Degraded vegetation visualized by orthophoto map.

3.1.2. Detailed Topographic Analysis and Geological Risk Refinement

The LiDAR data were pivotal for creating a high-precision Digital Terrain Model (DTM) with a 10 cm resolution, which, for the first time, revealed the detailed topography concealed beneath dense vegetation (Figure 8). This model enabled not just a visual assessment but the quantitative derivation of key morphometric parameters such as slope, aspect, and surface curvature (Figure 9). Based on calculations within the scanned area, the average terrain slope was determined to be 12.45 degrees, categorizing it as moderately sloped. Distinct sections with steep gradients exceeding 40 degrees are predominantly located between the developed areas and beachfront walkways of the surveyed territory. The highest observed slopes range between 50 and 60 degrees. By integrating the DTM with existing geological risk registries, the boundaries of active and potential landslide zones were delineated with sub-meter accuracy, significantly exceeding the precision of previously available maps. The analysis established a direct spatial correlation between unregulated pathways and parking areas on one hand, and the zones with the highest susceptibility to erosion on the other, thereby demonstrating the anthropogenic contribution to slope destabilization. The data were also applied in the design of drainage systems and other soil stabilization measures, permissible during the construction phase of the coastal park.

Figure 8. A precise digital terrain model obtained from aerial laser scanning.

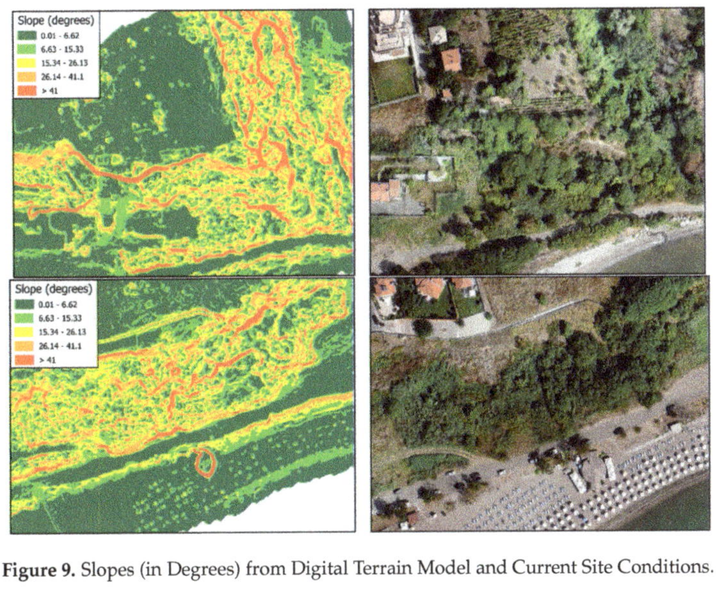

Figure 9. Slopes (in Degrees) from Digital Terrain Model and Current Site Conditions.

3.1.3. Three-Dimensional Structural Analysis of Vegetation

The combined use of terrestrial and airborne LiDAR data facilitated a shift from a traditional two-dimensional analysis of vegetation to a comprehensive three-dimensional assessment of its structure and architecture (Figure 10). Through point cloud segmentation, attribute data were extracted for individual trees, including their height, crown diameter, and overall volume. This approach allowed for the characterization of not only the areal extent but also the vertical complexity and density of the vegetation canopy. The analysis was further supported by field verification from landscape ecologists, allowing for the distinction between degraded areas needing restoration and those with naturally developed secondary vegetation. The latter, while performing a certain soil-stabilizing function, was characterized by low structural complexity and limited ecological value.

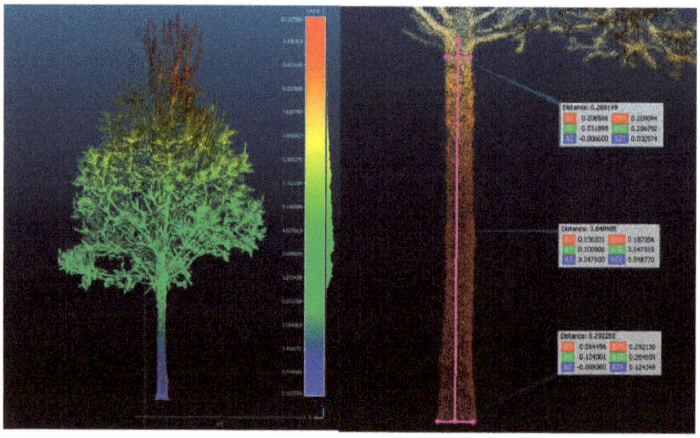

Figure 10. Segmentation and calculation of attribute data for individual trees.

The summary analysis of the above-mentioned newly generated data on the terrain and vegetation made it possible to calculate the carbon storage potential under different scenarios for the development of the territory. For the current situation, data on areas and types of vegetation (broadleaf, coniferous, shrub, and grass) were used, while for the future, data from the design team on those designated for removal and newly planned ones were used. The information is intended to support data-driven decision-making in the planning and design of the coastal area and the assessment of short- and long-term impacts.

3.2. Utilizing New Data for the Development of Coastal Development Scenarios

3.2.1. Scenario 0—Maintain the Existing Situation

The digital geospatial twin is applied as a corrective measure in analyzing the available information about the territory—Development zones according to the Master Plan of Burgas Municipality, Types of Ownership of properties, Way of permanent use of properties, Natura 2000 Protected Sites, Protected Areas, Types of Green Spaces, Transport Systems. Additionally, by comparing contour lines from 2009 and 2024 (via DTM), it identifies terrain dynamics caused by landslides, soil swelling, coastal erosion, and a wide range of anthropogenic interventions. A spatial analysis of the territory in the two land properties under consideration has been developed to assess the current situation. The findings indicate that approximately 29.52% of the area is currently deforested and disturbed (Figure 11). However, the existing vegetation, which mainly consists of deciduous trees and shrubs, plays a vital role in regulating coastal processes and influencing the local climate. Our calculations show a carbon sequestration rate of 24.3903 tC/ha/year.

Figure 11. Visualization of Scenario 0: The existing situation of the territory.

If the current situation and trends continue, we can expect deforestation to spread, fragmentation to increase, and biodiversity to be permanently damaged to the point of no return.

3.2.2. Scenario 1—"Construction of Parking Areas"

The scenario was developed in response to the identified shortage of parking spaces in the neighborhood. The pronounced elevation of the terrain is considered a limiting factor for pedestrian tourist traffic to the beach during the active summer season, which leads to a demand for parking spaces near the shore and, consequently, deforestation. The scenario is in line with the requirements of local businesses, beach concessionaires, and tourism representatives (hotels and restaurants). The case study area is not suitable for construction of underground parking facility due to a combination of geological risks, high groundwater levels, and the inability to support landscaping with tree vegetation on ground level. The total area designated for parking would amount to 21,947.89 m^2 (Figure 12), representing 15.56% of the park's total territory. According to the current Master plan /Land-use plan, the land plots are designated as "zones for landscaping, sports, and attractions" with an allowable building density parameter of up to 25%. Within these limits, the plan permits the construction of public service buildings and parking areas, and the paved surfaces of the alley network are not included in the calculation. The parking areas shown in the scenario remain below the allowable building density. No interventions are planned to improve the existing vegetation or enrich it further through afforestation. The zoning parameters outlined in the Master plan for the property permit a maximum building density of 25%. This allows for the construction of a building or parking lot with a height of up to 7 m, while also requiring a minimum of 25% green space, without altering the intended use of the area. Hypothetically, Scenario 1 complies with these parameters; however, a comprehensive analysis of the results from the digital twin indicates that it does not support a sustainable ecosystem or a favorable ecological status for the park. Investing in parking areas would meet the expectations of local business stakeholders and partially enhance visitation. This will lead to disruptions of the green space, impacting both its sustainability and environmental quality. Implementing this scenario would reduce the current carbon storage in the waterfront area's green spaces by approximately 22%. The construction of numerous parking spaces would provide only a partial solution within a limited residential area. In addition, opportunities for mitigation and adaptation to climate change are limited, and geological risk is significantly exacerbated by the high proportion of sealed soils and the load on the ground.

Figure 12. Scenario 1: Construction of parking areas—type of usability.

3.2.3. Scenario 2—"Enhancing Green Areas"

The scenario is based on the identified deforested areas, considering them as priorities for intervention (Figure 13). It represents a balanced option, including a set of measures for coastal improvement through: (1) Reinforcing degraded terrain, clearing degraded vegetation, and replanting to stimulate regulating ecosystem services—microclimate, surface water collection and drainage, and groundwater drainage, reinforcement and maintenance of soil structure; (2) Providing a network of paths with minimal slopes for pedestrians and cyclists, adapted to the terrain conditions; (3) Concentrating the construction of parking areas on the northern border of the properties and completely removing traffic from the landscaped area; (4) Protection, restoration, and enhancement of the natural environment and creation of conditions for recreational, sports, and educational activities.

Figure 13. Scenario 2: Increasing the green areas.

In the considered scenario, the parking areas amount to 4998.66 m^2, of which 3869.28 m^2 are located in deforested areas of the park and represent 2.74% of its total area. These values are well below the allowable building density of 25% according to the Master plan. The areas designated for re-greening amount to 37,754.12 m^2, representing 26.78% of the park's total area.

The impacts of this scenario cannot be considered in the short term due to the need for specific weather conditions for planting new vegetation and the several years required for strengthening, growth, and the commencement of ecosystem services. In the long term, the total carbon storage from increasing biomass = 11.4248 tC/ha/yr. This outcome is expected to enhance the carbon storage values of the current landscaped area of the coastal zone by nearly 27.8%.

3.2.4. Application of Scenarios for the Subsequent Design of the Coastal Park

Based on the data and analyses from the digital geospatial twin, the scenarios developed, and the expected short-term and long-term impacts, we can summarize that Scenario 2 reveals several advantages that align with our goals for an effective and climate-neutral revaluation of the territory in response to public expectations. Future development will focus on this to achieve and exceed the expected results. After the project implementation, the quantity of green areas will exceed 29 m^2/p, satisfying around 40% of the residential areas within a 300 m distance.

The conceptual design for the Coastal Park is based on the Vision: Preservation, restoration and emphasis of valuable natural features, sustainable management of geological risk, improvement of the connectivity of green and infrastructure in the conditions mitigation and adaptation to climate changes, and increasing the quality of the urban environment.

Special emphasis is purposefully placed on the structuring and functional organization of the park with an attitude towards naturalness and aesthetics, with details from local culture, ecological sustainability, and full-fledged regulating ecosystem services, as well as achieving year-round attendance and a high degree of accessibility for all people.

The alley network is planned parallel to the horizontals of the terrain to achieve minimal interference with the topography with natural, soft forms. The predominant longitudinal slope does not exceed up to 5% (Figure 14). This value is based on national legislation (REGULATION No. RD-02-20-2 of 20 December 2017, on the planning and design of the communication and transport system of urbanized areas (For bicycle lanes—Annex 8 to Article 62); REGULATION No. 2 of 29 June 2004 on the planning and design of communication and transport systems in urban areas—For pedestrian paths, Art. 115 (8) and For bicycle lanes, Art. 119 (1)), which does not specify exact parameters for pedestrian areas but sets requirements for bicycle lanes. Pedestrian and bicycle traffic is encouraged, with automobile traffic being taken outside the boundaries of the Coastal Park. Due to the presence of public service facilities at the beach, Sarafovo Fisherman's Port, and the Canal Pumping Station, it is necessary to provide conditions for the passage of service vehicles only at certain time intervals, as well as special regime/police vehicles, emergency medical assistance, and fire.

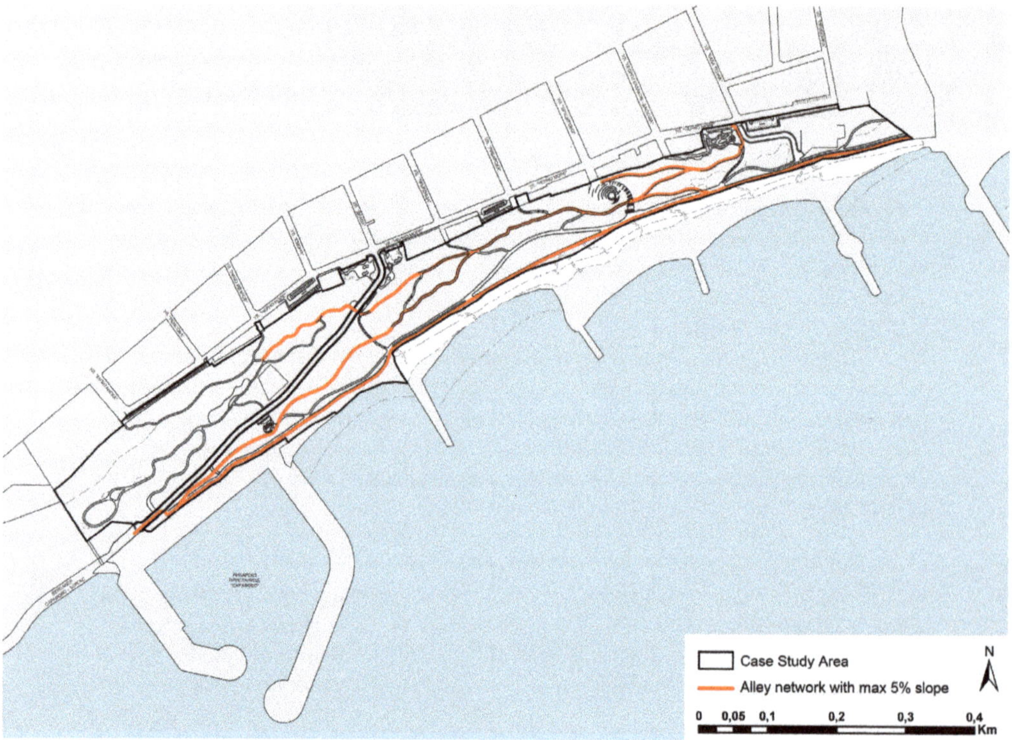

Figure 14. Accessible routes.

Multiple thematic zones have been identified, based on the proposals received in the Impact Model Workshop and field activities with participants from the Re-Value project, and further developed by planners and designers. Among them are playgrounds with thematic Black Sea and forest elements in the areas with the corresponding ecosystems, Panoramic terrace, Amphitheatre, Rope park and rope bridge for children and adults, Zones for prolonged stay and social interaction, Zones for sports—pump track and bike skills, Museum and park "Navigation" for representing the local life, communities, and culture, Dog park, and many more (Figure 15). All of them help to create a recognizable and functional coastal park, recognized by local communities and visitors, and emphasizing the coastal, forest, and urban ecosystem along with local culture.

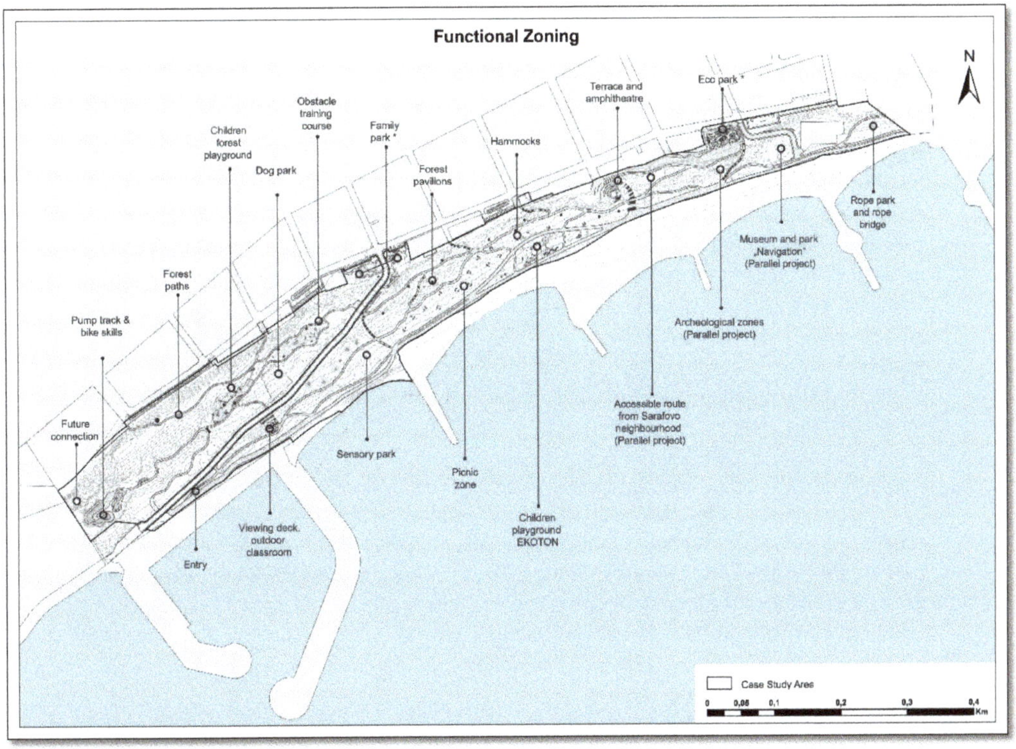

Figure 15. Functional Zoning.

Planting new vegetation can be carried out with the support and involvement of local communities through a series of events dedicated to the reforestation of selected species in specifically designated locations. A key outcome of this process could be the strengthening of community bonds around a shared, socially significant goal, as well as improved attitudes towards the vegetation and its future preservation.

4. Discussion

The developed digital geospatial twin of the Sarafovo district is an integrated digital replica of the physical territory, created through a combination of high-tech geospatial data and scientific methods. In the course of this study, it is of fundamental importance to overcome differences in data about the territory (scales, periods, methods of collection and organization, tools, accuracy, and purpose) and to combine the information into a single, precise model of the territory. This approach creates an information environment for active and productive cooperation between all actors involved in the planning and implementation of territorial interventions. In the context of the project, this digital model is a key tool.

4.1. Precise Mapping of Risks (e.g., Active Landslides, Erosion Zones) Based on Real Geological and Topographic Data

Monitoring inherited natural hazards in urban environments, such as landslides, is a serious challenge for the Bulgarian Black Sea coast. The Ministry of Regional Development and Public Works of the Republic of Bulgaria maintains structures for analyzing and mapping geological risks. These are represented as polygons or points with associated attribute data for registration code, risk levels, and affected area [39,40]. The digital geospatial twin provides a realistic, up-to-date picture of the risk circumstances and allows changes over time to be tracked. At the same time, this is an expensive solution for individual projects, and the rational approach would be for municipal structures to maintain and periodically update a digital twin of the city. This would enable the systematic tracking and reporting of current and future interventions. It would also provide a basis for comparing the initial, expected, and achieved results, as well as their impacts on a broader territorial scale. The recommended update intervals depend on the type of elements being monitored—annual updates are appropriate for technical infrastructure and construction, while a five-year cycle is more suitable for vegetation. In the context of the current territory and the conditions of a changing geography, such comparisons would be particularly valuable for geological risk management, afforestation success, and carbon sequestration, both within the site itself and as part of the broader neighborhood and citywide context. Digital twins support not only initial analysis and monitoring but also timely decision-making for interventions and the achievement of higher long-term outcomes.

4.2. Reassessment of the Advantages and Limitations Derived from Topography

Along with the vulnerability analysis, the identification of local topographical characteristics provided information for refining the alley network and thematic zoning of the park area so that the indicators relevant to such a project could be taken into account in the configuration of the landscape [18] concerning the quality of the coastal environment and public perception [41]—landscape aesthetic with a variety of natural and cultural elements for sports, recreation, cultural and educational activities, panoramic views, types of flooring and plant compositions.

4.3. Contemporary Vegetation Cover

The full utilization of urban forest functions, including their role as nature-based solutions, directly depends on object-oriented information and data quality as a basis for interdisciplinary collaboration [19]. The precise data generated in the course of this study greatly assisted in diagnosing the current state of the vegetation cover, both independently and in the context of the complex terrain. In some cases, the precise depiction within a digital twin necessitates data collection at different times of the year, directly addressing the varying phenological stages of trees. When detailed data on individual tree structure are required, optimal acquisition occurs during winter when deciduous trees lack foliage, enabling active sensors to extensively scan and map intricate tree parts. Conversely, for accurate canopy volume data, such as for biomass estimations, it is paramount to collect information during the period of maximum leaf development, ensuring the full extent of the tree crown is captured. The carbon footprint analysis carried out here was used as an argument in the discussion of development scenarios and in support of the final decision on the details of the urban park design. Here, we again defend the thesis that the full potential of the digital geospatial twin can be realized by discussing ecological phenomena on various scales. The current park design can be viewed in the context of the overall green-blue infrastructure of the city of Burgas. Such a spatial analysis would highlight the role of the park in carbon sequestration and the discussion of locations and types of new green infrastructure based on the ecosystem services demand [42].

Development and implementation of a Coastal Park concept will contribute to improving access to green areas for the residents and visitors of the Sarafovo quarter. However, in spatial terms, they do not provide a comprehensive solution, and in the future, other solutions should be sought for the establishment of small urban gardens in connection with green corridors to support their functionality and effectiveness. The project considers the implementation of green corridors to ensure the connectivity between coastal green spaces and those within the urbanized areas. However, such corridors are extremely limited due to the nature of land ownership, spatial configuration, and the high density of existing development.

Analyses based on the digital geospatial twin helped track the changing conditions in the development of the Sarafovo district, so that the new landscaping would stimulate the sustainability of the entire urban structure [43] and become the focus of the future construction of a unified green system for the district. The study of the structure of contemporary vegetation has been taken into account in the specific selection of plant species for new landscaping, structural combinations, and location. The species are both suitable for urban park design, representative of local biodiversity and at the same time ecologically resilient to the current and expected effects of climate change, to avoid costly challenges in the management of the park over time [44].

4.4. Visualization and Collaboration—3D Models Serve for Public Consultation by Clearly Demonstrating the Effects of the Scenario Proposals

Experience from contemporary practices shows that sustainability planning processes that use a geodesign approach can be improved with the help of more data-driven inputs and research on the usefulness of integrating data-based modeling and simulation into a collaborative scenario planning process [25]. The visualization of the territory, the subject of discussion and planning of interventions, is a very powerful tool regarding deeper engagement and better understanding of the needs and desires of, as well as a user-friendly sharing platform for, local stakeholders, residents, and visitors, allowing us to build a shared understanding of geodesign principles and apply them to a real-world problems. In our case, this is most clear in terms of demonstrating the high vulnerability of the terrain to active landslide processes and deformations of the modern coastal profile. The latter served as a strong argument in support of the construction of a coastal park, including in the direction of expanding the vegetation cover, zoning to achieve load balance, and isolation of vulnerable areas, as well as a reasoned concentration of critically needed parking areas in locations outside the park space.

The presented digital geospatial twin was not used in the discussion of engineering scenarios regarding traffic and other necessary socio-economic factors. Simultaneously, the digital twin emphasizes the most critical elements related to the functioning of the coastal strip we focus on—the coastal road, the port, and the internal road network in the neighborhood. Discussing these elements alongside other available information on daily travel and traffic congestion helped shape the final decision to suggest parking areas and to expand the intervention zone by including the current beach strip within the urban coastal park project area. Results from our experimental research on climate simulations are not presented here, which, given the proven predisposition of the territory to drought, the urban heat island effect, and flash floods during the period of winter precipitation, is very necessary to improve awareness in urban planning. Based on our previous research in the Burgas area [34], we consider this a necessity, which should, however, be carried out through a digital geospatial twin of the entire territory of the city of Burgas, which, in combination with appropriate data, e.g., thermal photogrammetry [34], would provide data to develop a general vision for the adaptation of the urban structure to climate change. We entirely share the understanding that adaptation measures for coastal cities should be designed with even greater attention to future hazards, exposure, and vulnerability [17], to promote the speed of their transformative adaptation [12].

The development of a high-fidelity digital twin for the Sarafovo district was accompanied by significant operational, technical, and resource-related challenges. The primary operational constraint was the study area's location within the controlled airspace of Burgas Airport, which necessitated a complex coordination and approval process with air traffic control authorities, restricting flights to narrow, pre-approved temporal and spatial windows. On the contrary, in the design process, the proximity to the airport does not constitute a constraint, as the project does not include any buildings or elements subject to height restrictions. Furthermore, the dynamic urban environment itself presented logistical hurdles; pedestrian and vehicle traffic required data acquisition to be scheduled during periods of low activity to minimize occlusions and motion artifacts. From a technical standpoint, the integration of heterogeneous datasets from three different sensor types was a major challenge, demanding meticulous post-processing to ensure seamless alignment and manage varying error characteristics and point densities. Environmental factors, such as inconsistent lighting affecting the radiometric quality of the photogrammetry, required careful planning and subsequent manual data cleaning. The manual editing aimed to mitigate inherent sensor limitations and rectify automatic processing anomalies, striving to meet a target positional accuracy typically within 10 cm for the point cloud data, consistent with high-resolution geospatial applications. Any potential deviation from this stringent quality control, driven by inconsistencies in manual intervention, directly influences the dataset's accuracy by introducing localized inaccuracies in feature representation and measurements. For example, if unmitigated positioning errors are allowed to persist, the digital twin would be severely hindered from being accurately integrated with other crucial datasets, such as cadastral data or other foundational spatial data, compromising its utility for comprehensive urban planning.

Finally, the project faced considerable resource challenges, not only due to the high cost of specialized hardware and software but also the immense volume of data generated, which demanded significant computational power and extensive, labor-intensive manual intervention for classification, refinement, and quality assurance.

5. Conclusions

The digital twins are a scientific and practical bridge between data and solutions. It transforms raw measurements into evidence for action, enabling planning that considers both environmental challenges and local community needs. This makes it indispensable for sustainable development in complex urban and natural environments. On the other hand, climate change is a fact, and it is obvious that it will have an increasingly negative impact on all geosystems that support life and balance on our planet. It will undoubtedly have an increasingly negative impact on urbanized areas, with cities that do not create conditions for the effective adaptation of their systems to these changes and their impact on the geographical environment being particularly affected. It should be borne in mind that this adaptation is a long-term and systematic process that must take into account the serious inertia of territorial development. Most cities rely on infrastructure systems that were designed and built under different conditions and are based on data that no longer accurately reflect the characteristics of the local climate and conditions, which puts them at serious risk. In this regard, modern geoinformation technologies, including digital geospatial twins, are among the key tools supporting adaptation processes in cities, both by providing an adequate basis for effective planning and management and for carrying out the necessary monitoring and response to changes in the geographical environment and the associated risks and threats.

A digital twin of a coastal residential neighborhood has been created to facilitate specialized geodesign and discussions with various stakeholders regarding three intervention scenarios in the coastal zone. The ultimate decisions focus on redesigning and constructing a coastal urban park that addresses critical environmental challenges in the area—such as active landslide processes and climate vulnerability—while also enhancing recreational opportunities and expanding public access to the coastline. The digital twin has been instrumental in integrating information from diverse sources into a cohesive model, significantly aiding interdisciplinary collaboration and public discussions during the reassessment of the coastline and the redesign of the urban park.

The present study shows only some of the possibilities of these technologies and tools. They clearly demonstrate both their serious potential to provide adequate information resources to support spatial planning processes and their capacity to integrate with other technological capabilities and systems towards the creation of "smart" and adaptive cities. The applied possibilities of these methods and tools, integrated into the concept of digital geospatial twins, are a "natural" solution that will define the framework for operational and strategic planning and management of urban systems in the near future. Today, spatial planning and management are carried out in a context of increasing "saturation" of all procedures with technological innovations. For this to happen in Bulgaria, it is necessary to encourage local authorities to implement these solutions, both by providing funds for technological improvement and increasing their technological capacity, and by working to improve knowledge and skills [45].

Finally, it should be borne in mind that adaptation to climate change is a global policy with specific local dimensions. Unlike the typical mechanism for structuring and shaping policy instruments, climate change adaptation does not draw on past experience and data but relies on complex simulations and models to provide the information base on which it can be developed.

Author Contributions: Conceptualization, S.D., B.B. and L.S.; methodology, B.B., A.I. and L.S.; software, M.I. and S.D.; formal analysis, B.B., A.I., M.I. and L.S.; writing—original draft preparation, B.B., S.D., A.I. and L.S.; writing—review and editing, B.B., S.D., A.I., M.I. and L.S.; visualization, M.I., A.I. and S.D.; supervision, S.D. and M.R.; project administration, M.R. and Z.S.; funding acquisition, M.R. All authors have read and agreed to the published version of the manuscript.

Funding: This study is financed by the European Union-NextGenerationEU, through the National Recovery and Resilience Plan of the Republic of Bulgaria, project No. BG-RRP-2.004-0008-C01 "SUMMIT—Sofia University Marking Momentum for Innovation and Technological Transfer".

Data Availability Statement: The original contributions presented in the study are included in the article, further inquiries can be directed to the corresponding author.

Acknowledgments: We are very grateful for the support from the project ReValue, Horizon Europe Grant Agreement No: 101096943. We also thank the reviewers and the academic editors for their valuable comments that helped to improve the paper's quality.

Conflicts of Interest: The authors declare that the research was conducted in the absence of any commercial or financial relationships that could be construed as potential conflicts of interest.

References

1. Elmqvist, T.; Andersson, E.; McPhearson, T.; Bai, X.; Bettencourt, L.; Brondizio, E.; Colding, J.; Daily, G.; Folke, C.; Grimm, N.; et al. Urbanization in and for the Anthropocene. *Npj Urban Sustain.* **2021**, *1*, 6. [CrossRef]
2. Olivadese, M.; Dindo, M.L. Water, Ecosystem Services, and Urban Green Spaces in the Anthropocene. *Land* **2024**, *13*, 1948. [CrossRef]

3. Jiang, B.; Yao, X. (Eds.) Geospatial Analysis and Modeling of Urban Structure and Dynamics: An Overview. In *Geospatial Analysis and Modelling of Urban Structure and Dynamics*; Springer: Dordrecht, The Netherlands, 2010; Volume 99. [CrossRef]
4. Jin, A.; Ge, Y.; Zhang, S. Spatial Characteristics of Multidimensional Urban Vitality and Its Impact Mechanisms by the Built Environment. *Land* **2024**, *13*, 991. [CrossRef]
5. Łaszkiewicz, E.; Wolff, M.; Andersson, E.; Kronenberg, J.; Barton, D.N.; Haase, D.; Langemeyer, J.; Baró, F.; McPhearson, T. Greenery in urban morphology: A comparative analysis of differences in urban green space accessibility for various urban structures across European cities. *Ecol. Soc.* **2022**, *27*, 22. [CrossRef]
6. Steenberg, J.W.N.; Millward, A.A.; Nowak, D.J.; Robinson, P.J.; Ellis, A. Forecasting Urban Forest Ecosystem Structure, Function, and Vulnerability. *Environ. Manag.* **2017**, *59*, 373–392. [CrossRef]
7. Miao, C.; Wang, J.; Wang, D. Research progress on urban forest ecosystem services and multifunctionality. *Int. J. Environ. Sci. Technol.* **2024**, *22*, 11557–11566. [CrossRef]
8. Parsa, V.A.; Salehi, E.; Yavari, A.R.; van Bodegom, P.M. Evaluating the potential contribution of urban ecosystem service to climate change mitigation. *Urban Ecosyst.* **2019**, *22*, 989–1006. [CrossRef]
9. Sanesi, G.; Gallis, C.; Kasperidus, H.D. Urban Forests and Their Ecosystem Services in Relation to Human Health. In *Forests, Trees and Human Health*; Nilsson, K., Sangster, L., Gallis, C., Hartig, T., Seeland, K., Schipperijn, J., Eds.; Springer: Dordrecht, The Netherlands, 2011. [CrossRef]
10. Konijnendijk, C.C.; Annerstedt, M.; Nielsen, A.B.; Maruthaveeran, S. Benefits of urban parks: A systematic review. *A Rep. IPFRA.* 2013.
11. Pinto, L.V.; Inácio, M.; Ferreira, C.S.S.; Ferreira, A.D.; Pereira, P. Ecosystem services and well-being dimensions related to urban green spaces—A systematic review. *Sustain. Cities Soc.* **2022**, *85*, 104072. [CrossRef]
12. Wannewitz, M.; Ajibade, I.; Mach, K.J.; Magnan, A.; Petzold, J.; Reckien, D.; Ulibarri, N.; Agopian, A.; Chalastani, V.I.; Hawxwell, T.; et al. Progress and gaps in climate change adaptation in coastal cities across the globe. *Nat. Cities* **2024**, *1*, 610–619. [CrossRef]
13. Shah, M.A.R.; Van Dau, Q.; Wang, X. Social-ecological vulnerability and risk to coastal flooding and erosion in major coastal cities. *Int. J. Disaster Risk Reduct.* **2025**, *118*, 105286. [CrossRef]
14. Cosby, A.G.; Lebakula, V.; Smith, C.N.; Wanik, D.W.; Bergene, K.; Rose, A.N.; Swanson, D.; Bloom, D.E. Accelerating growth of human coastal populations at the global and continent levels: 2000–2018. *Sci. Rep.* **2024**, *14*, 22489. [CrossRef]
15. Smith, H.; Ferrari, M.S. (Eds.) *Waterfront Regeneration: Experiences in City-Building*; Routledge: Oxfordshire, UK, 2010.
16. Wei, Z.; Ahern, J. Novel Urban Waterfront Ecosystem Services Evaluation, Monitoring and Improving Strategies. *Fábos Conf. Landsc. Greenway Plan.* **2016**, *5*. [CrossRef]
17. Le, T.D.N. Climate change adaptation in coastal cities of developing countries: Characterizing types of vulnerability and adaptation options. *Mitig. Adapt. Strateg. Glob. Change* **2020**, *25*, 739–761. [CrossRef]
18. Du, H.; Zhou, F.; Fu, Y. Impact of Urban Waterfront Green Spaces on Public Health and Well-Being Based on User Perspectives. *J. Urban Plan. Dev.* **2025**, *151*, 04025001. [CrossRef]
19. Zhao, J.; Davies, C.; Veal, C.; Xu, C.; Zhang, X.; Yu, F. Review on the Application of Nature-Based Solutions in Urban Forest Planning and Sustainable Management. *Forests* **2024**, *15*, 727. [CrossRef]
20. Blaschke, T.; Hay, G.J.; Weng, Q.; Resch, B. Collective Sensing: Integrating Geospatial Technologies to Understand Urban Systems—An Overview. *Remote. Sens.* **2011**, *3*, 1743–1776. [CrossRef]
21. Ricciardi, G.; Callegari, G. Digital Twins for Climate-Neutral and Resilient Cities. State of the Art and Future Development as Tools to Support Urban Decision-Making. In *Technological Imagination in the Green and Digital Transition*; The Urban Book Series; Arbizzani, E., Cangelli, E., Clemente, C., Cumo, F., Giofrè, F., Giovenale, A.M., Palme, M., Paris, S., Eds.; Springer: Cham, Switzerland, 2022. [CrossRef]
22. Ranatunga, S.; Ødegård, R.S.; Onstein, E.; Jetlund, K. Digital Twins for Geospatial Decision-making. *ISPRS Ann. Photogramm. Remote Sens. Spat. Inf. Sci.* **2024**, *X-4-2024*, 271–278. [CrossRef]
23. Bai, H.; Li, Z.; Guo, H.; Chen, H.; Luo, P. Urban Green Space Planning Based on Remote Sensing and Geographic Information Systems. *Remote Sens.* **2022**, *14*, 4213. [CrossRef]
24. Zellner, M.L.; Massey, D.; Laboy, M.; O'Brien, D.T.; Mueller, A.; Engelberg, D. Enhancing digital twin technology with community-led, science-driven participatory modeling: A case in green infrastructure planning. *Environ. Plan. B Urban Anal. City Sci.* **2025**, 23998083251323671. [CrossRef]
25. Debnath, R.; Pettit, C.; Leao, S.Z. Geodesign Approaches to City Resilience Planning: A Systematic Review. *Sustainability* **2022**, *14*, 938. [CrossRef]
26. Hayek, U.W.; von Wirth, T.; Neuenschwander, N.; Grêt-Regamey, A. Organizing and facilitating Geodesign processes: Integrating tools into collaborative design processes for urban transformation. *Landsc. Urban Plan.* **2016**, *156*, 59–70. [CrossRef]
27. Amoroso, N.; Pierre, A.; Olson, L.; Pearsell, G.; Lavigne, D. Geodesign: Applications towards Smarter Planning and Urban Design Solutions. *J. Agric. Environ. Sci.* **2019**, *8*, 4–13. [CrossRef]
28. Steinitz, C. *A Framework for Landscape Planning Practice and Education*; Process Architecture: Tokyo, Japan, 1990.

29. Steinitz, C. *A Framework for Geodesign: Changing Geography by Design*; Esri Press: Redlands, CA, USA, 2012.
30. Shannon, M.; Rouse, D. *Geodesign and the Future of Planning*; American Planning Association: Chicago, IL, USA, 2015; pp. 1–10.
31. Miller, W.R. Introducing Geodesign: The Concept. 2012. Available online: https://www.esri.com/content/dam/esrisites/sitecore-archive/Files/Pdfs/library/whitepapers/pdfs/introducing-geodesign.pdf (accessed on 27 June 2025).
32. National Statistical Institute, Republic of Bulgaria, Population Data from 31 December 2022. Available online: https://www.nsi.bg/en (accessed on 27 June 2025).
33. Plan for Integrated Development of Burgas Municipality 2021–2027. Available online: https://plan.smartburgas.eu/%D0%BF%D0%B8%D1%80%D0%BE-2021-2027/ (accessed on 27 June 2025).
34. Dimitrov, S.; Borisova, B.; Iliev, M.; Semerdzhieva, L.; Themistocleous, K.; Michaelides, S.; Hadjimitsis, D.G.; Papadavid, G. Application of thermal photogrammetry in local climate zones definition for urban heat island effect assessment and mapping of the city of Burgas, Bulgaria. In Proceedings of the Ninth International Conference on Remote Sensing and Geoinformation of the Environment (RSCy2023), Ayia Napa, Cyprus, 21 September 2023. [CrossRef]
35. Methodological Guidelines for the Preparation of Consolidated Documentation for Demonstrating Climate Resilience (Including Climate Change Mitigation and Adaptation to Climate Change) of Project Proposals Under the Operational Programme. Environment 2021–2027 of the Republic of Bulgaria. Available online: https://www.eufunds.bg/en/node/12996 (accessed on 27 June 2025).
36. Task force on National Greenhouse Gas Inventories, IPCC. Available online: https://www.ipcc-nggip.iges.or.jp/public/2006gl/pdf/4_Volume4/V4_04_Ch4_Forest_Land.pdf (accessed on 27 June 2025).
37. EIB Project Carbon Footprint Methodologies, European Investment Bank, January 2023. Available online: https://www.eib.org/attachments/lucalli/eib_project_carbon_footprint_methodologies_2023_en.pdf (accessed on 27 June 2025).
38. Bulgaria's National Greenhouse Gas Inventory Report, Executive Environment Agency, Republic of Bulgaria. Available online: https://old.eea.government.bg/en/output/unfccc/index.html (accessed on 27 June 2025).
39. Analysis, Assessment and Mapping of Geological Risk at the Ministry of Regional Development and Public Works. Available online: http://gis.mrrb.government.bg/KGR/ (accessed on 27 June 2025).
40. Geoprotection Companies at the Ministry of Regional Development and Public Works. Available online: https://geozashtita.bg/ (accessed on 27 June 2025).
41. Lyu, M.; Wang, S.; Shi, J.; Sun, D.; Cong, K.; Tian, Y. Visual Satisfaction of Urban Park Waterfront Environment and Its Landscape Element Characteristics. *Water* **2025**, *17*, 772. [CrossRef]
42. Menconi, M.E.; Bonciarelli, L.; Grohmann, D. Assessment of the Potential Contribution of the Urban Green System to the Carbon Balance of Cities. *Environments* **2024**, *11*, 98. [CrossRef]
43. Alberti, M. Cities of the Anthropocene: Urban sustainability in an eco-evolutionary perspective. *Philos. Trans. R. Soc.* **2024**, *379*, B37920220264. [CrossRef]
44. Esperon-Rodriguez, M.; Gallagher, R.; Calfapietra, C.; Cariñanos, P.; Dobbs, C.; Eleuterio, A.A.; Rodriguez, D.E.; Jahani, A.; Litvak, E.; Livesley, S.J.; et al. Barriers and opportunities for resilient and sustainable urban forests. *Nat. Cities* **2025**, *2*, 290–298. [CrossRef]
45. Batty, M. *The New Science of Cities*; MIT Press: Cambridge, MA, USA, 2013; ISBN 9780262019521.

MDPI AG
Grosspeteranlage 5
4052 Basel
Switzerland
Tel.: +41 61 683 77 34

Land Editorial Office
E-mail: land@mdpi.com
www.mdpi.com/journal/land